Metamodeling-Driven IP Reuse for SoC Integration and Microprocessor Design

For a listing of recent related Artech House titles,
turn to the back of this book.

Metamodeling-Driven IP Reuse for SoC Integration and Microprocessor Design

Deepak A. Mathaikutty
Sandeep K. Shukla

ARTECH HOUSE
BOSTON | LONDON
artechhouse.com

Library of Congress Cataloging-in-Publication Data
A catalog record for this book is available from the U.S. Library of Congress.

British Library Cataloguing in Publication Data
A catalogue record for this book is available from the British Library.

ISBN-13 978-1-59693-424-5

Cover design by Yekaterina Ratner

685 Canton Street
Norwood, MA 02062

10 9 8 7 6 5 4 3 2 1

Contents

Preface

“We recognize IP reuse as the key to creating reliable, proven, repeatable design content on a truly industrial scale” [1]. “When it comes to SoC design, nothing is more important in enabling productivity, reducing risk, and simplifying verification than the use of standardized, preengineered blocks and subsystems” [2]. These two quotes, one from the president of the SPIRIT consortium, and another from a regular contributor to the *IC Design and Verification* online journal, summarizes the mantra that also led us to embark on a project resulting in this book. Today in 2009, almost everyone in the semiconductor industry recognizes that reuse is the key to higher designer productivity and reduced time to market. The complexity of hardware systems has been increasing at a rapid pace due to the rising expectations by consumers to use sophisticated applications. At the same time, due to the immense competitive pressure, the companies are hard pressed to reduce the design turn-around time. As a result, the semiconductor and the embedded systems industries have been seeking the Holy Grail or a panacea for almost a decade now. At this point in time, IP reuse and its enablers for efficient reuse are pitched as the solution everyone has been seeking. We concur with many industry and academic experts: it might very well be a prudent direction to take, if not the only one.

However, reuse is not a new phenomenon in the industry. Cut-and-paste reuse is common in most of the software and hardware design industries. In software engineering, the technology has moved beyond the cut-and-paste reuse soon after the inception of procedural programming. Soon after, the object-based and object-oriented programming concepts made their way into the standard software design paradigms. Component based design and reuse are essential elements of software design for decades now. In fact, one almost cannot even write any useful C program without using standard library function calls. This is not to claim that software engineering has solved all its problems with productivity and correctness. We just mention this in order to provide a historical context of reuse as a technological precept.

In contrast, even in the early 1990s, hardware design at the register-transfer-level (RTL) or below did not evolve any reuse methodology beyond some proprietary coding standards, and some methodological rigors. The reusability manual [3] was the closest to the best advice a hardware designer could get at that time. In our personal experiences in the industry, we have seen rampant use of cut-and-paste reuse from one generation of a microprocessor to the next at a semiconductor company. This practice was common to the extent that even the names of variables and wires gave away the fact that some of the code were cut and pasted from previous generation RTL. This obviously caused verification problems to no end. The verification effort spent in finding the problems originating from such unstructured reuse often outspent the time saved in reusing existing code base. Clearly, reuse as a methodology, and further more, reuse as a technology were sorely needed.

Before, we discuss more about the reuse as technology which is the main theme of this book; let us recapitulate a little bit of some recent history of design automation to ease our way into more discussion on IP reuse technology and its benefits.

Today, it is sort of well accepted that IP reuse might be our best shot at substantial productivity gain. However, in the late 1990s, a lot of hopes and resources were invested into the idea of moving beyond the RTL. The fact that RTL replaced schematics based design entry a decade earlier, and the success of logic synthesis in the intervening years provided the historical perspective along these lines. The experts advocated creating high-level design languages, and high-level synthesis from functional models to RTL. It is necessary that such hopes are kept alive. In fact, we are hopeful that languages that abstract away from the RTL details would become the design-entry language in the future. Unfortunately, the last 10 years have evolved rather differently than the course charted by the high-level modeling aspirations in 1999.

For a while, it was believed that C++ and object oriented programming would make expressing design intents easier. The added advantage would be the free compilation path leading to free and fast simulation for functional testing. In September 1999, Synopsys announced the release of SystemC-0.9 library for hardware design. Contemporaneously, Cynapps (which has been renamed Forte Design Systems) released Cynlib. The designers who believed that moving to higher abstraction level is the key to higher productivity were excited. The released versions of these libraries implemented class libraries that can model most structural RTL level constructs including modules, architectures, clocks, and so on. The concurrency was modeled with threads. However, it was soon realized that the simulation efficiency can only be gained if the model of computation that drives the simulation is of higher granularity than the ones employed in discrete-event simulation.

Most designers who were experts in using languages such as VHDL (originally invented as a specification and later simulation language) and Verilog used these C++ libraries very much like structural RTL languages. The simulation speed-up expected did not materialize in the experiments. The notion of transaction level modeling started taking shape. Various ideas were added to SystemC library in the version 2.0 onwards, from another academic language, namely SpecC. The notions of channels, events, and various methods of channel synchronizations were added. Over the years, due to the strong industrial involvement in the SystemC initiative, various levels of transactional abstractions have been proposed, and various add-on libraries have been implemented. A number of academic researchers did not think that transactions are the only way to make modeling easier and closer to designer intent. They advocated behavioral modeling through addition of models of computation specific constructs, implementing the simulation of these constructs through redesign of the kernel [4, 5] or additional libraries [6]. Over time, a new industrial buzzword sprang into existence: ESL (Electronic System Level) languages.

What ESL really entails has many interpretations, and many volumes have been written [7] about the ESL methodologies and languages. Despite their interpretational differences, the common understanding seems to be that SystemC at the right level of abstraction is a representative of ESL, and so is SystemVerilog. Various other languages claim to be ESL, and some argue in favor and some against. The point is that ESL has not yet proven to be the all encompassing solution to the designer productivity problem. Neither is IP reuse. However, there seems to be a confluence of opinions there. The technologies and standards that enable IP reuse is certainly required by the industry. The IP design houses as well as users of predesigned IP blocks all agree on this. Along came the SPIRIT consortium in 2001.

SPIRIT [1] consortium aims to standardize the language and tool independent exchange format for IPs. They also proposed various tools (called generators) that could use such exchange format descriptions to build various collaterals including verification models and other aspects of a purchased IP. At the time of the birth of this consortium, another industry consortium called the VSIA (Virtual Socket Interface Alliance) already worked for several years on standardization efforts for IP interfaces. They published a number of standards for IP exchange formats and various bus interface standards, to make System-on-Chip design more efficient. A few other consortiums or joint industry efforts also looked into standardizing various on-chip bus interfaces at various abstraction levels, such as AMBA and OCP.

Without going into too much details about the history of IP interfacc standardization efforts, it is sufficient to say that IP reuse needs these standardizations, and it must be successful for the productivity gain in a world where IP reuse and IP exchange based design of systems will become the trend.

In the above rendition of our view on a piece of industrial history, we tried to convince the readers that IP reuse will have an immediate impact on designer productivity than ESL alone. The IP reuse standards and technologies enabling IP reuse are not the main topic of this book, but it is highly related. In fact, our effort in creating enablers for IP reuse happened in parallel with efforts such as those of VSIA or SPIRIT. However, we must embrace the standards, but the technology enablers can differ in the way they work. This book describes how we thought to create such technologies, often unaware of the standardization efforts in progress elsewhere.

Now, let us go through a brief history of how we commenced on the path that led to the research presented in this book. We believe that the history I am going to narrate now will have relevance to the readers interested in our views and technologies for IP reuse.

In 2000, Sandeep Shukla met Professor Rajesh Gupta while working as a design automation engineer in the semiconductor industry. This was the time when we were experimenting with SystemC-0.9 and Cynlib-1.0 to prove or disprove that modeling with these libraries would make simulation faster than standard HDL languages. Since Rajesh worked on SystemC when it was code named Scenic, before its open source release, it was natural to seek his advice. During one such conversation, he mentioned that a framework for composing preexisting components would be the right way to solve the productivity problem. At the same time, he was embarking on a project with a graduate student Frederic Doucet and Masato Otsuka, a visitor from a Japanese company at his lab at the University of California at Irvine. They were using Tcl/Tk to create a scripting language that would allow describing a skeleton of an SoC architecture. This schematic architecture would then be instantiated by querying a database for existing components. They called this project BALBOA. The idea quite was fascinating, and in April 2001, Sandeep Shukla quit his industrial job to join Rajesh Gupta's lab as a postdoctoral researcher to help on the BALBOA [8–12] project.

It was immediately recognized that to solve the problem of composition of components through a Component Composition Framework (CCF), we needed to do the following: (1) expose the interfaces of the components using some kind of metadata or interface description language (IDL), (2) define a metalanguage to describe the skeletal architecture of the SoC (visual or textual language), (3) implement reflection/introspection technique so that the interface information for all existing components are exposed to the algorithms that match the skeletal component holders to real components based on interface types, (4) define a type-system for the interfaces and implement type matching algorithm, and (5) implement test generation methodology to test the composed system.

In [8] the original Tcl/Tk based CCF of BALBOA was introduced along with the BALBOA Interface Description Language (BIDL). In [12] we described the introspective features and reflection mechanisms; in [9] we described the interface data types, the type matching algorithm and its complexity. In [11, 12] various other aspects of this system were described in detail. In 2002, we came to another conclusion through our experiments for building example SoCs with BALBOA. We needed to expose more than the interface port data types and limited behavioral information exposed through BIDL.

Along with Jean-Pierre Talpin and his group in INRIA, France, we started looking into behavioral types for the interfaces to allow assume-guarantee reasoning for the validating that the composed components behave correctly. This led to several theoretical results, mostly about using a polychronous [13] language to expose some behavioral aspects of the signals at the interface of the components from the library, and the corresponding type theory and inference algorithms. Automated methods for exposing these behavioral descriptions were proposed. This was done by constructing SIGNAL [14] programs from the static single assignments views during the compilation of SystemC components. Various aspects of this work were published in [15–17].

At this point, we got initiated into two of the latest technologies: first, we started using XML [18] schema for describing interface types in a format that is independent of the language in which the components are written. Second, we started using a wonderful metamodeling environment for creating domain specific design languages. This environment allows the use of such newly defined languages to create domain specific models. The metamodeling rules and constraints allow modeling-time dynamic check on the models' compatibility with the language rules. This tool called GME (Generic Modeling Environment) [19] came out of Janos Sztipanovits's group at Vanderbilt University. These two technologies changed the way we were building the CCF in the BALBOA project. We first tested the idea of defining our own interface description schema in XML, and used GME in building an environment for designing embedded software from functions written in a functional language (SML). This framework code named EWD [20] led us onto the next CCF and in this book we describe this CCF in detail. We call this CCF, the metamodeling-based component composition framework (MCF).

Thanks to the use of GME, MCF allowed us to build a visual language to describe SoC architectures as schematics. We could even use aspect views of the same model to expose various aspects of the model separately, leading to the use of the principle of separation of concern. We could also build languages for transactional model composition as well as RTL model composition. We could add syntactical elements to describe all kinds of design components such as buses, point-to-point interconnects, and network-on-chips. We could provide syntactic wizards for instantiating simple RT level components such as mux/demux, as well as functional/algorithmic components. This way the MCF system evolved into a comprehensive CCF that we originally wanted to build with the BALBOA project. The credit of building this system in less than 2.5 years goes to Dr. Deepak Mathaikutty. I must mention two other features that we had to add to make MCF a full-fledged IP reuse enabler.

First, in BALBOA we used BIDL to describe the interface of components. An IP designer has to supply the BIDL along with the component itself for proper use of the component in the CCF. In that sense, BIDL was a precursor of IPXACT. Earlier, in connection with another project, Deepak worked with two other students, Hiren Patel and David Berner, who together built a parser for SystemC along with some post processing (called SystemCXML) [21]. This parser allowed us to extricate the structural information for an arbitrary SystemC model, similar to those manually put in the BIDL files. They even created a graphic viewer, which can visualize an arbitrary SystemC model with its hierarchical structure, ports, and connections. However, SystemCXML could not extract much behavioral information from the components. As a result, Deepak built another parser on top of another SystemC parser (KarSCPar) from Karlsruhe University. This new parser allowed extraction of more information, which was used for reflection/introspection purposes. For any arbitrary preexisting SystemC model, this new parser could extract enough information that could later be used during component selection, matching types, and creating PSL properties for testing of component composition.

Second, the behavioral type theory that we developed in [14–16] was problematic to implement because that would have required generation of SIGNAL programs as behavioral types from the SystemC models. The tools for such extraction were being developed at INRIA, and we did not want to wait for the delivery of that tool. There were a few delays due to change in personnel in that project. Therefore, the component compatibility checking through type inference was getting delayed. This posed a challenge to us, like any other software engineering project. However, in order not to delay the construction and demonstration of the CCF, we formulated another way of solving the checking of compatibility of components while choosing them for composition. The interface data extracted with Deepak's new parser had enough information to build temporal properties about the components' behaviors. As a result, a mechanism to test composition of various possible alternative components was automated by Deepak. Of course, unlike type theoretic algorithms, such tests did not necessarily formally prove compatibility. This provided higher confidence while selecting components from the library to instantiate an SoC architectural schema. As a proof of concept for demonstrating an IP reuse enabler based on metadata, and behavioral compatibility checking, this was enough. In the future, however, we expect to integrate the tool for SIGNAL based behavioral type extraction and type matching algorithms for full formal guarantee of compatibility.

The other aspect of this book deals with using similar ideas to build verification IPs, and that led to our collaboration with Intel and University of Minnesota to a parallel framework called MMV. During one summer internship at Intel, the first author showed MCF to the validation engineers. Dr. Ajit Dingankar drafted him to work over an entire semester, along with a visiting student from Minnesota to build the MMV environment.

It is noteworthy that our work on using XML schema to define language for IP interface specifications were done before the IPXACT was released by the SPIRIT consortium. Of course, our schema does not match with the IPXACT due to several reasons. First, we were not aware of the SPIRIT efforts until the 2006 Design Automation Conference; second, IPXACT was defined with tool integration in mind, so that specific IP contents can be kept proprietary while exposing enough information for exchanging IP to other tool environments and facilitating the IP reuse in a different company's tool flow. While we share the need for exposing the IP interface information, our sole aim has been enabling component composition automatically. Certain intellectual property protection issues were not our concern at the time.

IPXACT 1.4 was released in 2008 along with two generators which can build various views of an IP. Proprietary generators can also be plugged in through APIs defined by SPIRIT consortium. The scope of IPXACT is much grander. The effort involved more than 70 companies with dedicated engineering resources. The work described in this book is based on the Ph.D. work of one graduate student. However, as SPIRIT releases IPXACT and further standardizations, we feel that researchers interested in such metadata based IP reuse ideas could benefit from our approach. We believe that there are valuable lessons to learn beyond the specifics of our XML schemas or the specific architecture of our tool. The entire methodology and algorithms for matching components against their placeholders in the architectural schema, we hope, will prove uscful to engineers, researchers, and managers interested in IP reuse based design projects.

Most of the work in this project was funded by an NSF CAREER grant, an SRC Integrated Systems Funding, and an industrial funding from the Intel Corporation.

References

[1] SPIRIT, "The Spirit Consortium Press Release," Munich (DATE 2008), March 10, 2008.

[2] K. Morris, "The Spirit of Standardization: IP Re-Use Takes Center Stage," *IC Design and Verification Journal*, Techfocus Media Inc., April 2006.

[3] M. Keating and P. Bricaud, *Reuse Methodology Manual for System-on-a-Chip Designs*, 3rd Edition, Norwell, MA: Kluwer Academic Publishers, 1998.

[4] H. D. Patel and S. K. Shukla, *Systemc Kernel Extensions for Heterogeneous System Modeling: A Framework for Multi-MoC Modeling and Simulation*, Boston, MA: Kluwer Academic Publishers, 2004.

[5] H. D. Patel and S. K. Shukla, *Ingredients for Successful System Level Automation and Design Methodology*, Boston, MA: Springer, 2008.

[6] F. Herrera and E. Villar, "A Framework for Heterogeneous Specification and Design of Electronic Embedded Systems in SystemC," *ACM Transactions on Design Automation of Electronic Systems*, Vol. 12(3), No. 22, April 2008, pp. 1–31.

[7] T. Grotker, S. Liao, G. Martin, and S. Swan, *System Design with SystemC*, Boston, MA: Kluwer Academic Publishers, 2001.

[8] F. Doucet, S. Shukla, M. Otsuka, and R. Gupta, "An Environment for Dynamic Component Composition for Efficient Co-Design," *Proceedings of the Conference on Design, Automation and Test in Europe (DATE'02)*, 2002, pp. 115–122.

[9] F. Doucet, S. Shukla, M. Otsuka, and R. Gupta, "BALBOA: A Component Based Design Environment for System Models," *IEEE Transactions on Computer Aided Design of Integrated Circuits and Systems*, Vol. 22, No. 12, December 2003, pp. 1597–1612.

[10] F. Doucet, S. Shukla, and R. Gupta, "Introspection in System-Level Language Frameworks: Meta-Level vs. Integrated," *Proceedings of the Conference on Design, Automation and Test in Europe (DATE'03)*, 2003.

[11] F. Doucet, S. Shukla, and R. Gupta, "Structured Composition Techniques for Embedded Systems," *Proceedings of International Conference on High Performance Computing (HiPC)*, Springer Verlag, December 2002.

[12] F. Doucet, R. Gupta, M. Otsuka, S. Shukla, and P. Schaumont, "Interoperability as a Design Issue in C++ Based Modeling Environments," *Proceedings of the 4th International Symposium on System Synthesis (ISSS'01)*, 2001, pp. 87–92.

[13] Polychrony, Project-Espresso: The Polychronous Approach Web site, http://www.irisa.fr/espresso/Polychrony/.

[14] P. L. Guernic, J. P. Talpin, and J. C. Lann, "Polychrony for System Design," *Journal of Circuits, Systems and Computers, Special Issue on Application Specific Hardware Design*, April 2003.

[15] J. P. Talpin, D. Berner, S. K. Shukla, P. L. Guernic, A. Gamati, and R. Gupta, "A Behavioral Type Inference System for Compositional System-on-Chip Design," *Proceedings of the International Conference on Application of Concurrency to System Design (ACSD'04)*, Hamilton, Canada, IEEE Computer Society, June 2004.

[16] D. Berner, J. P. Talpin, P. L. Guernic, and S. K. Shukla, "Modular Design Through Component Abstraction," *Proceedings of the International Conference on Compilers, Architecture, and Synthesis for Embedded Systems (CASES'04)*, September 2004, pp. 202–211.

[17] J. P. Talpin, P. L. Guernic, S. K. Shukla, and R. Gupta, "A Compositional Behavioral Modeling Framework for Embedded System Design and Conformance Checking," *International Journal of Parallel Programming*, Vol. 33, No. 6, June 2005, pp. 613–643.

[18] W3C Consortium, W3C Web site, http://www.w3.org/.

[19] GME, GME Web site, http://www.isis.vanderbilt.edu/projects/gme/.

[20] D. A. Mathaikutty, H. Patel, S. Shukla, and A. Jantsch, "EWD: A Metamodeling Driven Customizable Multi-MoC System Modeling Framework," *ACM Transactions on Design Automation of Electronic Systems*, Vol. 12(3), No. 33, 2007, pp. 1–43.

[21] H. D. Patel, D. A. Mathaikutty, D. Berner, and S. K. Shukla, "CARH: A Service Oriented Architecture for Validating System Level Designs," *IEEE Transactions on Computer-Aided Design of Integrated Circuits and Systems*, Vol. 25, No. 8, August 2006, pp. 1458–1474.

Acknowledgments

The first author would like to thank Dr. Ajit Dingankar for his guidance and support. The first author would also like to acknowledge the support received from the NSF CAREER grant CCF-0237947, an SRC Integrated Systems Grant, and industrial grant from the INTEL Corporation, which provided the funding for the work reported in this book.

Special thanks are due to Deepak's roommates and friends from and around FERMAT as well as Intel: Syed Suhaib, Gaurav Singh, Debayan Bhaduri, Hiren Patel, Sumit Ahuja, Bijoy Jose, Nimal Lobo, David Berner, Animesh Patcha, Manoj Selvatharasu, Harish Barathvajasankar and Sourabh Joshi.

Deepak thanks his colleague and friend Sreekumar Kodakara for introducing the author to the domain of microprocessor validation and working with him on many different Intel projects.

Deepak thanks his best friend and wife Leena Jacob for her inspiration, support, and love through the many hardships during his graduate studies. She is his better half and without her, the last four years of his graduate life would have been very lonely. She always manages to make his plain life more colorful with her simple gestures, cute flights, funny expressions, and finally her beautiful smile.

Deepak thanks his childhood friend Sunish John for his belief in him and faith in his success. To all the other friends that have come and gone, thank you all.

Finally, a heartfelt thanks goes to his parents Manattu Abraham Mathaikutty and Sosamma Mathaikutty, his in-laws Kuruvilla Jacob and Beena Jacob, Leejo Jacob and Libin Jacob, his sister Deepa Niju and her husband Niju Aelias and their son Nikhil Niju, his brother Dony Mathaikutty, his cousins Anish J. Mathew and Anila J. Mathew, and his grandmother Sosamma Abraham for their continued love, support, and encouragement in his endeavors.

Chapter 1

Introduction

Due to the rapid advances in silicon manufacturing technologies, silicon capacity has been steadily doubling every 18 months as predicted by Moore's Law [1]. Figure 1.1 shows the trend in silicon manufacturing using data collected from one of the major microprocessor vendors [2]. The continual technological advances allow companies to build more complex systems on a single silicon chip, both System-On-a-Chip (SoC) as well as custom processors. However, their ability to develop such complex systems in a reasonable amount of time is diminishing with the increase in complexity [3]. This is referred to as the *productivity gap* [4] as shown in Figure 1.2, which is based on the ITRS (International Technology Roadmap for Semiconductors) [4], where the gap between plot P_1 indicating the advances in engineering productivity and plot P_2 representing the silicon complexity is widening.

As the complexity and functionality of electronic systems increase, the possible design choices and the alternatives to explore for optimization purposes also increase. Therefore, design space exploration (DSE) is vital when constructing a system in order to choose the optimal alternative with respect to power, area, and performance. However, DSE is a time-consuming and tedious part of the system design process. The ability to reduce the time to develop these system-level models for optimization purposes can dramatically improve design quality and accelerate design exploration. A possible way to reduce this time is to reuse preexisting cores, which are implementations of various standard blocks [5]. Such previously implemented blocks are often called intellectual property (IP) cores.

In this book, we address two related but distinctive problems faced by today's system design companies, namely reusing design IP for fast SoC integration, and reusing verification IP for generation of tools for microprocessor validation.

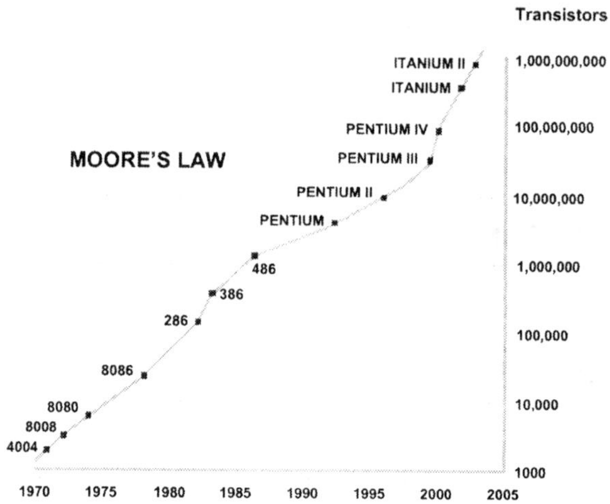

Figure 1.1 Moore's Law.

Accordingly, the book is divided into Part I and Part II, which are related but describe the two problems and our solutions to them. Therefore, we consider two types of IPs: (1) design IP and (2) verification IP.

Design IP

Many small and large design teams use third-party IPs to accelerate the SoC design process by importing functional blocks, such as data converters, PLLs, DSPs, processors, and memory, which are examples of design IPs. In other words, predesigned building blocks that designers can just drop into their flow to perform a predefined function are called *design IPs*. Databahn [6] is an example of a high-speed memory controller IP core from Denali Software, Inc. [7].

Verification IP

Like design IPs, verification IPs can also, in theory, be designed for reuse or licensed from third parties. Verification IPs are developed to become a part of the testbench or verification environment and used to ensure the correctness of the design. They do not become a part of the design itself. Commercial verification IPs often only consist of a bus-functional model (BFM) with a monitor to do some protocol checking. An example of a verification IP is DW_IBM440 [8] from Synopsys Inc. [9] for the powerPC 440 microprocessor core.

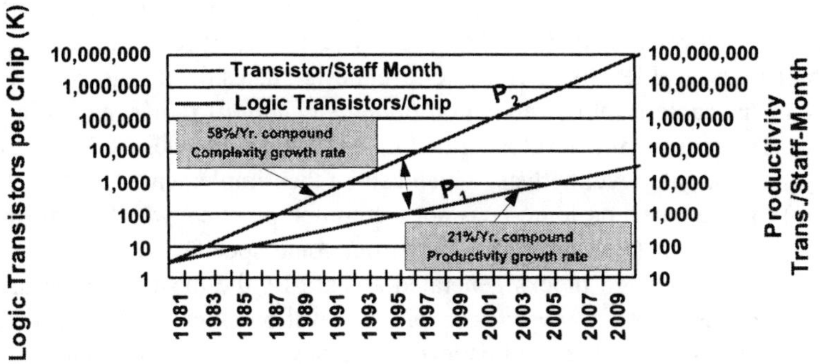

Figure 1.2 Productivity gap.

Design Reuse

Design reuse is defined as the inclusion of previously designed components (design IPs) in software and hardware. Design reuse makes it faster and cheaper to design and build a new product, because the reused components will not only be already designed but also tested or verified for correctness. Hence, it is a necessity for efficient implementation of SoC designs. According to a recent study conducted, design teams can save an average of 62% [10] of the effort that they would otherwise expend in the absence of reusing previously designed blocks. Reuse of IPs span across the different abstractions, which requires extending the IP definition and reuse to different levels of abstraction. We briefly summarize the characteristics of IPs at each abstraction level in the following paragraph.

Table 1.1

Comparison of Hard-IP Versus Soft-IP

IP	Representation	Optimization	Technology	Reuse
Hard IP	GDSII	Very high	Dependent	Low
Soft IP	RTL	Low	Independent	Very high

At the lowest level of abstraction the design is modeled using a circuit description. It is specified as a netlist describing the connections between components such as resistors and transistors. More abstract physical devices such as ideal voltage or current sources are also specified as a part of the circuit description. The IP-reuse model at this level is called *hard-IP* [10], where hard-IP refers to a technology specific layout level IP block. The gate-level design takes the design complexity to a more manageable level by relying on *Boolean equations* rather than voltage and currents. The reuse at this level is termed *logic-IP*. This involves exchange of technology specific gate-level netlists, which can be then implemented by the physical design. IP reuse based on logic-IP is not a common practice, which paves the way for register transfer (RT) level IP. Apart from some specific blocks like memory or highly critical design blocks generated as hard-IP, the most common usage of the term IP reuse refers to a *synthesizable RT* level block called *soft-IP* [11]. The basic unit of design modeled at this level is a clock cycle. The RTL model defines the microarchitecture of a design in terms of its internal states and the behavior between the cycles of the design. The next natural extension of IP reuse is called *arch-IP*, where the basic entity modeled is a *system task*. The model at this level specifies the computational or communication processes and the parallelism that exists in the design. The final classification is termed as *functional-IP*, which is an important aspect for SoC design. At this level we are dealing with *functions* of the system. A clear interface definition and development for functional-IP can expedite IP reuse through proper selection and well-defined integration of a system for a complex SoC.

Design reuse based on soft-IP, arch-IP, and functional-IP has been in use for quite some time with limited success in certain domains [11]. The success of the reuse methodology requires analyzing the nature of design style and scope of reuse at RTL and higher levels of abstraction. Table 1.1 summarizes the comparison of the two prominent IP classes based on their features from [12].

Verification Reuse

A challenge with the increasing design complexity is the high cost associated with fixing a bug in a released product, which makes verification a key ingredient in the product development cycle. For example, consider microprocessors that are modeled at different levels of abstraction as shown in Figure 1.3. The complexity of models increases as we go down the abstraction hierarchy. Verification complexity at the microarchitecture level of abstraction dwarfs the complexity at higher levels of abstraction because of the additional features introduced in the microarchitecture to support performance optimizations such as out-of-order, superscalar, and speculative execution to name a few. Hence, currently most of the verification effort is spent in validating the functionality introduced in the microarchitecture and implemented at the RTL. This trend may change in the future generations of multicore processors that are designed with multiple cores in a single die and with hardware support for software applications like virtualization [13], transactional memory [14], and XML parsing [15]. This shift in design paradigm of pushing all the features on to the processor increases the verification complexity at *all* levels of abstraction of the processor. Therefore, for products such as microprocessors where the complexity and hence the cost of verifying a microprocessor increases from one generation to the next, verification is an important and challenging problem for current and future designs. Hence, the designers and validators employ various tools and techniques to ease and quicken the verification task. Examples of such tools are simulators, test generators, coverage metrics and test plans, which are collectively called *verification collaterals*. A possible approach to enabling reuse in the verification task is to employ the models created for the various processor abstractions to derive the different verification collaterals. A model that is reused to systematically derive the various verification tools and targets is called a *generative verification IP*. The idea behind verification IP reuse is "model once and reuse many times." This is a *model-driven* approach towards a verification strategy. The alternative would be to write testbenches, coverage monitors, simulators, and so forth, individually, which is tedious, error-prone and incurs many more person-hours.

1.1 ONGOING EFFORTS IN DESIGN IP REUSE

The VSI Alliance (VSIA) [16] specifies "open" interface standards and specifications that facilitate the integration of software and hardware virtual components from multiple sources for the development of SoCs. The phrase *virtual component* is used to describe reusable IP components that are composed with other components (as if plugging them onto virtual sockets), similar to how real hardware components are plugged into real sockets on a board. VSIA eases the integration work required to incorporate IP-cores into "virtual sockets" at both the functional level (e.g., interface protocols) and the physical level (e.g., clock, test, and power structures). This will allow IP providers to productize and maintain a uniform set of IP deliverables, rather than requiring them to support numerous sets of deliverables needed for the many unique customer design flows.

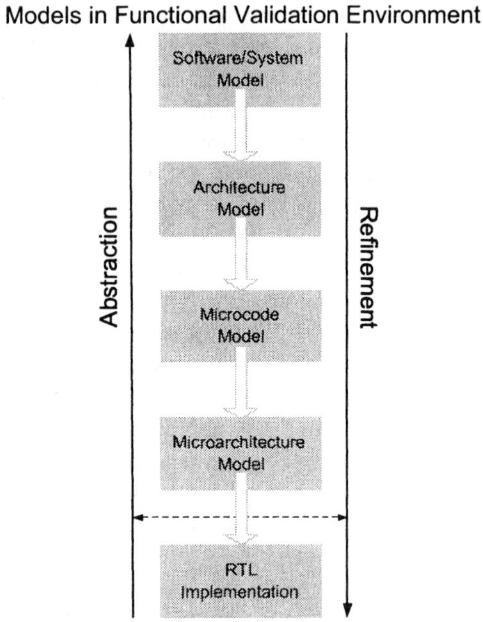

Figure 1.3 Different levels of abstraction in a microprocessor.

The OCP International Partnership Association, Inc. (OCP-IP) [17] promotes and supports the open core protocol (OCP) as the complete socket standard that ensures rapid creation and integration of interoperable virtual components. OCP-IP offers a universal socket standard with communication socket for an IP to specify data flow signaling, control signals, and test signals. It shields IP designers from the knowledge of the interconnection system, and enables the same IP to be portable among different systems.

An orthogonal approach to reusing IP-cores is through a design flow that utilizes metadata [18] exchanged in a design-language neutral format, as proposed in SPIRIT [19]. It proposes an IP-reuse approach to enable IP configuration and integration by utilizing metadata exchanged in XML [20] format between tools and SoC design flows. SPIRIT provides the API standards that should be implemented to export the design and configure the actual IPs to create the executable. However, the designer is required to specify the IP-explicit metainformation, which is a time-consuming task and affects the essential rapid DSE.

For communication performance analysis during IP integration, the GreenSocs initiative developed the GreenBus [21, 22]. This SystemC fabric simulates the behavior of both on-chip and interchip communication buses and networks. GreenBus can be configured to arbitrary (user-defined) communication protocols and provides different levels of abstraction, thus facilitating early embedded software development and cycle-count accurate communication architecture exploration.

Despite all these efforts and support from the semiconductor industry, there are still many challenges to successful design IP reuse. All of the above reusability standards ease the compatibility issues between IP-cores to an extent. On the other hand, none of them with perhaps the exception of the GreenBus exactly goes close to solving the engineering problem of quick construction of system-level models from reusable IPs, so DSE can be expedited. However, in the case of the GreenBus, the architecture assumes that all IP components communicate over a bus, which forces the DSE to incur the overhead associated with bus protocols. Therefore, we need a framework that allows the designer to architect the system in terms of components and their composition called a *component composition framework* (CCF).

In this book, we present such a design framework called metamodeling-driven component composition framework (MCF), which allows the designer to be a non-expert on the specifics of the software language and enables realizing a system through *design templates* and *tool support.*

The proliferation of SystemC [23] IP-cores necessitates CAD frameworks that support IP-core composition to build system models by reducing much of the software engineering (SE) and C++ programming burdens that a designer has to face in creating models from existing cores. Examples of such frameworks are BALBOA [24] and MCF [26]. The BALBOA [24] composition environment relies on "smart wrappers" that contain information about the types and object models of the components, and assemble them through an interpreted environment with a loose typing. The particularity of the environment is the split-level programming with type abstraction and inference [25]. We developed MCF [26], a metamodeling [27, 28] based component composition framework for SystemC based IP-core composition. It provides a tool for automated selection [18] and substitution of IPs in an architectural template [29], making use of metadata synthesized from an IP library [30].

1.2 ONGOING EFFORTS IN VERIFICATION IP REUSE

Synopsys [9] provides a wide range of verification IPs for the industry's most popular bus protocols such as AMBA 3 AXI, AMBA 2.0, PCI Express, USB 2.0 OTG, Ethernet, Serial ATA, and thousands of memory models. Their product DesignWare Verification IP [31] can be integrated with testbenches written in SystemVerilog [32], Verilog [33], VHDL [34], and OpenVera [35], to generate bus traffic, to check for protocol violations, and to monitor and generate extensive reports on functional coverage of the bus protocols. Their support for the SPIRIT XML data model [19] is enabling them to provide SPIRIT views of some of their verification IPs for the purpose of reuse [36].

Incisive Verification IP [37] from Cadence [38] also provides a solution for the common protocols used in today's applications. These include PCI Express, AMBA AHB, and AXI, Ethernet, USB, PCI, OCP, and SATA. This product offers verification modules called *universal verification components* (UVCs) for testbenches, transaction-based acceleration, assertion-based verification, emulation rate adapters, and in-circuit verification. Their UVCs can be integrated with testbenches written in languages such as SystemVerilog [32], e [39], Verilog [33], VHDL [34], and SystemC [23]. The latest addition to the incisive verification product is a graphical tool called scenario builder [40], which can be used for creating reusable stimulus and test cases on top of an existing Specman Elite [41] environment. The user selects from the reusable data collection to create useful scenarios for the verification task.

An SoC verification environment called FlexBench [42] provides a multilayer architecture to enable reuse of verification IPs. The lowest layer (stimulus layer) provides the user-interface for a design-under-test as well as all the drivers and monitors. The highest layer (integration layer) includes all of the drivers and monitors in the simulation. The middle layer, called the service layer, connects the other two layers and provides some system services.

Despite all these efforts and support from the electronic design automation (EDA) industry, there are still many challenges to successful verification IP reuse. All of the above EDA verification tools and IPs improve the verification experience of the user and reduce the productivity gap to an extent. On the other hand, none of them solve the problem of enabling effective reuse of reusable verification components, which requires highly reconfigurable IPs and robust verification capability such as directed random stimulus generation, protocol and temporal checking, and structural and functional coverage metrics. Furthermore, none of them specifically address reuse of verification IPs to enable a "model once, use many times" verification strategy, which is essential for a product such as a microprocessor, where the verification complexity spans across abstraction levels. Therefore, we need a unified framework to model the processor at different abstraction levels and then consistently generate various verification tools from all these models.

In this book, we present a validation environment called metamodeling-driven microprocessor validation environment (MMV), which allows the validator to describe relevant parts of the processor at different abstraction levels and then click a set of buttons to generate verification collaterals such as simulators, test generators, and coverage monitors.

1.3 ESSENTIAL ISSUES WITH IP REUSE

Reuse has created a whole new branch of semiconductor industry called *IP business* [43]. From the discussions in the previous section, we see that IP basically is a know-how technique that can be in the form of algorithms, software, designs, models, test suites, or methodologies. This know-how can be modified and tuned to a particular application. Most designers in most companies use old know-hows in their new designs and products. Capturing this information, distributing, and reusing it represents the core of IP business.

Essential Issues with Design IP Reuse

IP reuse has three major players: (1) IP providers, (2) IP integrators, and (3) IP tool developers. The IP providers supply IP components, including hard cores, soft cores, and software components. IP integrators are typically designers in system companies. IP tool developers provide IP providers and integrators with design methodologies and tools to support IP development and system integration. In order to make IP reuse easy, we need to address the issues of these three groups [3,44,45].

(1) IP Provider

An IP provider works on capturing the information of the know-how, distributing and reusing it, thereby representing the core of IP business. In the IP provider domain, some of the important components are IP library, documentation, quality assurance, and standardization. Some of the essential issues with these components are discussed below.

IP Library

One of the problems faced by IP providers is the definition of an IP in terms of functionality, style, and type. If the IP definition is very specific then it is easier to optimize, verify, and test it. On the other hand, if the definition is too generic, then it is more difficult to make any assurance on the quality. Another problem is how to convert a legacy code or design into a reusable IP. Any possible solution must account for all design aspects from the point of view of IP integrators, such as expandability, testability, and migration.

Documentation

Organization of catalogs or databases for IP, and the format and content of IP information for query search and evaluation, is another issue for IP providers. A possible solution is to develop standards on IP information (metadata for integration and interoperability) and capture the information in a design-language neutral format that can be exchanged between IP providers and IP integrators.

Quality Assurance

One of the problems is that most soft-IPs from third-party vendors have not been verified in silicon. Hence, quality and performance are not well known. Another problem is that it is difficult to verify that the given IP is functionally correct and delivers what it promises in the specification for all values of all parameters. A possible solution to these problems is that the third-party vendors must have an IP qualification methodology [46] in place, to measure the quality of the IP and the generated documentation should be exchanged along with the IP.

Standardization

Each IP vendor may have their own specification and documentation format. In order to evaluate IPs from different vendors, the user may have to apply a series of transformation procedures. This is both tedious and time-consuming, which makes it a potential problem. Standardization of the specification style and format to capture the different design aspects of the IP such as interfaces and algorithms, will ease this problem. Examples of a few are OCP-IP [17], VSIA [16], and IP-XACT from SPIRIT [19] that provide standards for easing the compatibility issues between IP-cores to a certain extent.

(2) IP Integrator

IP integrators need a standardized IP evaluation mechanism to compare similar IP-cores from different sources and to access the risk involved with using a selected core in a specific chip. They are responsible for the integrity of the design developed by composing IPs, therefore they have to verify the IPs employed as well as their interconnection. In the IP integrator domain, some of the essential components and issues are discussed below.

Exploration

First, designers need to know what design alternatives exist and which IP-cores to select for those choices. Second, designers need to compare IPs provided by different vendors with similar functionality to select the ideal one for their design. Hence, *an easy-to-use design exploration methodology is required to facilitate the IP selection process for system integrators.*

Integration

System integration by reusing existing IP-cores is not a trivial task. To successfully reuse existing IPs, designers not only need to fully understand IP specifications but also need to rely on support provided by IP vendors. Some of the problems that arise from reusing IPs for system integration are type mismatch, incompatible interfaces both at the structural level (pin/port-level mismatch) as well as at the behavioral level (protocol mismatch or type incompatibility), and timing errors.

Methodology and Environment

Establishing the infrastructure for design-for-reuse is the most important first step for embracing the design-for-reuse culture. Furthermore, in today's design practice, most system companies are using an RTL-based design environment. This method is not sufficient to support IP reuse. In order to adopt IP-reuse strategies, integrators will have to move the design entry to higher levels of abstraction, which would allow performing architectural trade-offs, power optimizations, and so forth, earlier on in the design flow.

(3) Tool Developer for IP Reuse

The electronic design automation (EDA) industry plays a pivotal role in the success of IP reuse. The industry should develop tools that provide a systematic design methodology to support IP reuse. In the IP tool domain, the essential problems are discussed next.

Support for IP Provider

First, tools for easy technology migration of IPs are needed to keep up with the fast pace of technology advances. Also, a standard verification and testing methodology is required to improve the quality assurance of IPs. Second, tools for supporting documentation preparation and searching are very important.

Support for IP Integrator

First, tools and methodologies supporting high-level design abstraction beyond the RT level are needed to support DSE through IP reuse. Second, a design exploration and IP evaluation methodology and its supporting tools are needed to facilitate the design reuse. Third, a reuse-automation design methodology is needed to alleviate the tedious system integration process by automatic selection and composition of IPs to create efficient simulation models.

The requirements to ease the system integration through reuse of design IPs are targeted in Part I of this book.

Essential Issues with Verification IP Reuse

Just like design IP, verification IP needs to be a self-contained, interoperable, modular, and plug-n-play product. However, even more than design IP, verification IP must be user-friendly, easy to comprehend and operate, extremely configurable, and highly customizable. Good-quality verification IP offers tremendous value to design teams trying to reuse design IP-cores into their project, regardless of whether they are testing for compliance or verifying the system integration. However, for verification IP to be effective in enabling reuse of complex design IP, it needs to provide many more features and functions than just a bus functional model. In our work, we focus on enabling reuse of generative verification IPs for a microprocessor verification strategy. Some of the essential requirements to ease their reuse are summarized next.

(1) Modeling Language

First, a programming language-independent modeling syntax and semantics is needed to create models at the various modeling abstractions in a uniform way. Second, a way to express rules is needed that enforces well-defined model construction in each abstraction level and consistency across abstractions. Third, a visual editor is needed to facilitate the modeling using the specified syntax and semantics and enforce the modeling rules through checkers during design time. Finally, the language should be easily customizable by the language developer for future requirements.

(2) Generation Algorithms

Generation algorithms facilitate the automatic code generation of verification collaterals such as test plans, coverage monitors, and test plans, from the models. A wish list to enhance the usability and extendibility of such algorithms would include first, the capability to plug-in interpreters that perform model analysis, transformation, and translation into executables is needed and second, the capability to handle language extensions and remain independent of the implementation target specifics, which can be provided by the user during verification.

The requirements and wish list to ease the reuse of generative verification IPs are targeted in Part II of this book.

1.4 METAMODELING APPROACH TO REUSE

A *metamodel* [47] is a formal definition of the constructs, rules, and constraints needed for creating application-specific models. The analysis, construction, and development of a metamodel for an application domain is called *metamodeling* [48]. The language used to define the metamodel is called a *metamodeling language* [27]. A framework with the capability of creating metamodels is called a *metamodeling framework*. The generic modeling environment (GME) [49] is an example of such a metamodeling framework that allows the description of a modeling domain by capturing the domain-specific syntax and static semantics into visual notations and rules called the *metamodel*. The dynamic or execution semantics of the modeling domain are provided through interpreters that can be plugged into the framework. They analyze and translate the models into executables. The metamodeling-based approach allows users to capture the designer's intuition, knowledge, and design constraints specified as rules into an abstract notation (metamodel). It can also express the model of computation, which provides the execution semantics of the models being built. The metamodel generates a formal framework that enforces certain rules that the designer has to conform to during model construction, thereby providing a design approach that is *correct-by-construction*. In GME, the metamodeling capability is provided through a set of generic concepts implemented as class diagrams in the Unified Modeling Language (UML) [50] and as Boolean constraints in the Object Constraint Language (OCL) [51], resulting in visualization, design-time checks, and customization.

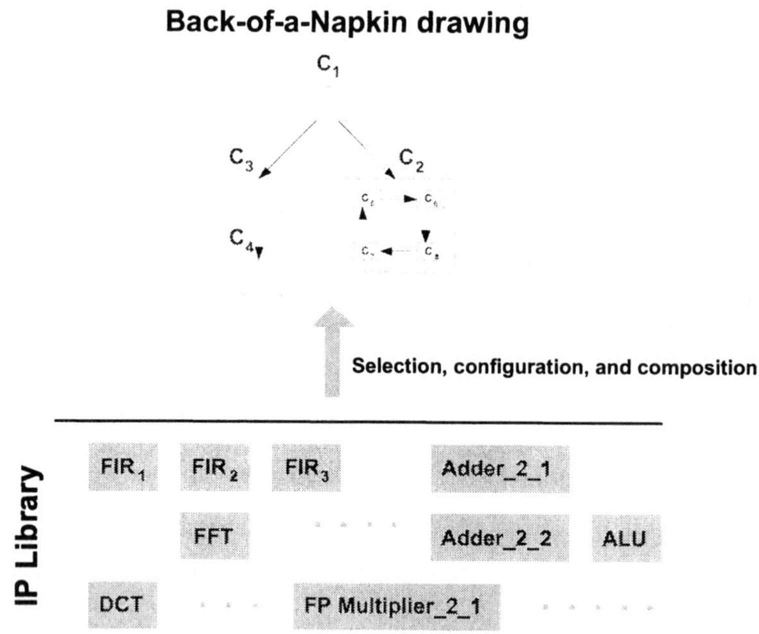

Figure 1.4 Design reuse.

Metamodeling Approach to Design IP Reuse

Our approach to design reuse is a component composition framework (CCF) [26] that makes the designer oblivious to the specifics of the software language and allows realizing a system through design templates and tool support. A design template is derived from the metamodel and it structurally reflects the modeling abstractions such as register-transfer level (RTL), transaction level (TL), mixed-abstractions (RTL-TL), and the structural hierarchy [28]. Furthermore, the metamodeling-driven modeling framework allows design visualization and enforces checks that look for inconsistencies, datatype mismatches, incompatible interfaces, and illegal component or channel descriptions [29]. The IP-reuse use-case model targeted by a CCF is shown in Figure 1.4, which is typically how designers go about designing. They sketch a *back-of-the-napkin drawing* that pictorially describes the various components of the design and their interaction. Furthermore, if design reuse is one of the objectives, then the library of IPs developed during the previous projects is searched and appropriate preexisting implementations are reused to instantiate the blocks in the diagram.

Metamodeling Approach to Verification IP Reuse

Our approach to generative verification IP reuse is a metamodeling-based modeling and validation environment [52] as shown in Figure 1.5. The various microprocessor models (architectural, microarchitectural) are uniformly described using a formal visual language. The modeling environment enables a stepwise manual refinement from one level to another and enforces consistency of the refinements. Finally, the environment facilitates automatic generation of verification collaterals such as simulators, coverage monitors, checkers, and test plans, for aiding the verification task of microprocessors at all levels of the modeling abstraction.

1.5 PROBLEM STATEMENT

The overarching problem is to enable design and verification for SoCs and microprocessors, which promote effective reuse of design IP as well as verification IP leveraging metamodeling techniques.

On the design front, we want to enable rapid design space exploration during the design of SoCs by providing languages, tools, and techniques for building fast simulation models by automating reuse from a library of SystemC IPs.

On the verification front, we want to enable verification of the architecture and microarchitecture implementations of a processor by providing languages, tools, and techniques for creating system-level models for the relevant parts of the processor and automatically generating verification collaterals such as functional simulators, test generators, and coverage constraints by reusing these models.

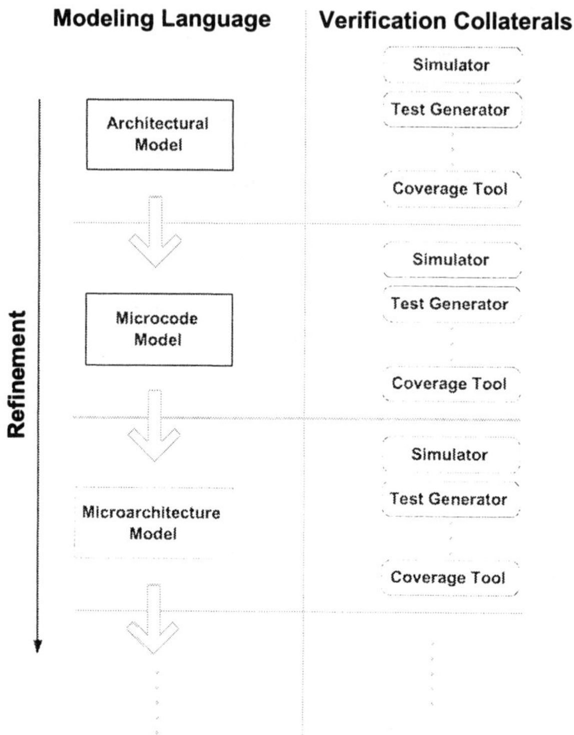

Figure 1.5 Reuse of verification IPs for collateral generation.

The problems pertaining to design IP reuse addressed in Part I of this book are:

1. How do you rapidly and precisely specify the SoC design requirements (structural, compositional, behavioral, and temporal)?
2. What is the IP metainformation relevant for enabling IP reuse? How is this metadata identified, extracted, and represented?
3. How do you quickly construct multiple simulation models by automating reuse from an IP library exploiting the metainformation for each IP?

4. How do you prove or automatically check the correctness of the composition of IP components?

The problems pertaining to verification IP reuse addressed in Part II of this book are:

1. How do you precisely model the different parts of the processor?
2. How do you capture the various modeling abstractions uniformly and enforce a well-defined refinement flow across abstractions?
3. What are the verification collateral that need to be generated from the processor models and how are they automatically generated?

1.6 RESEARCH CONTRIBUTIONS

The research contributions in Part I of this book for enabling design IP reuse are as follows:

1. Defined the metamodels, constraints and developed various checkers to enable a visual architectural specification of the design requirements in terms of components and their interaction [26, 29].
2. Developed reflection and introspection-based data mining techniques to extract structural and composition-related properties of a design IP, which was represented using XML schemas and data structures [30].
3. Formulated type theoretic algorithms for IP selection and composition to enable automated reuse from an IP library [18, 53].
4. Identified the type inference problem that exist in our IP-reuse solution and analyzed the complexities of the inference algorithms developed as part of our solution [54].
5. Developed property specification-based verification flow to enhance the correctness of the simulation models constructed [55].

The research contributions in Part II of this book for enabling verification IP reuse are as follows:

1. Defined a metamodeling-based language with the different abstraction levels to specify the relevant parts of the processor [52].

2. Developed a model-driven approach to automatically generate verification collaterals. The collaterals targeted in this book are functional simulator, random test generation, and coverage-directed test generation as well as design-fault directed test generation [56–58].

3. Presented the modified condition decision coverage (MCDC) [59] as a metric for microprocessor validation [60, 61].

4. Performed a comparative study and established a good correlation between the types of design faults seen during software validation [62] and the bugs reported in case studies on microprocessor validation [60].

1.7 TOOLS AND TECHNIQUES DEVELOPED

The frameworks developed in this book to enable IP reuse for design and verification of SoCs and microprocessors are as follows:

- Design IP reuse (Part I)
  1. Developed *MCF*, a visual language and framework for component composition using the metamodeling paradigm with design-time constraints to check for system-level inconsistencies, type mismatches, interface incompatibilities during modeling, and composition of models.
  2. Developed tool support for automated reflection, introspection, selection, and composition of IPs based on structural type theoretic principles from given IP libraries.
  3. Developed a property-based checker generation approach for enhancing the correctness of IP composition.
- Verification IP reuse (Part II)
  1. Developed *MMV*, a visual microprocessor modeling and validation environment using metamodeling techniques, which enables creating models of the system, architecture and microarchitecture of the processor.
  2. Developed a model-driven test generation framework that detects the design faults as well as performs coverage-directed test generation.

3. Developed an integrated framework for model-driven development and validation of system-level designs with a combination of Esterel and SystemC.

References

[1] G. E. Moore, "Cramming More Components onto Integrated Circuits," *IEEE Electronics*, Vol. 38, No. 8, 1965, pp. 114–117.

[2] The INTEL Corporation, http://www.intel.com/technology/mooreslaw/.

[3] D. D. Gajski, A. C.-H. Wu, V. Chaiyakul, S. Mori, T. Nukiyama, and P. Bricaud, "Essential Issues for IP Reuse," *Proceedings of Asia and South Pacific Design Automation Conference (ASPDAC'00)*, 2000, pp. 37–42.

[4] International Technology Roadmap for Semiconductors, http://www.itrs.net/.

[5] K. Werner, "IP Reuse Gets a Reality Check," *Chip Design*, June 2005.

[6] "Denali's Databahn™Memory Controller Cores Achieve Production/Gold Status in UMC's IP Program," D & R Industry Articles, http://www.us.design-reuse.com/news/news4321.html, November 2002.

[7] Denali Software, Inc., http://www.denali.com/.

[8] "Verification IP Catalog for Synopsys, Inc." http://www.synopsys.com/ipdirectory.

[9] Synopsys Inc., http://www.synopsys.com/.

[10] C. W. H. Strolenberg, "Hard IP Offers Some Hard-Core Values," D & R Industry Articles, http://www.us.design-reuse.com/articles/article2552.html, May 2000.

[11] R. Chandra, "IP Reuse and Platform Base Designs," D & R Industry Articles, http://www.us.design-reuse.com/articles/article6125.html, August 2003.

[12] J. Ebert and F. Nekoogar, *From ASICs to SOCs: A Practical Approach*, Upper Saddle River, NJ: Prentice Hall, 2003.

[13] R. Uhlig, G. Neiger, D. Rodgers, A. L. Santoni, F. C. Martins, A. V. Anderson, S. M. Bennett, A. Kagi, F. H. Leung, and L. Smith, "INTEL® Virtualization Technology," *Computer*, Vol. 38, 2005, pp. 48–56.

[14] T. Harris, S. Marlow, S. Peyton-Jones, and M. Herlihy, "Composable Memory Transactions," *Proceedings of the Tenth ACM SIGPLAN Symposium on Principles and Practice of Parallel Programming*, 2005, pp. 48–60.

[15] Layer 7 Technologies, "XML Accelerator," http://www.layer7tech.com/.

[16] Virtual Socket Interface Alliance, http://www.vsi.org.

[17] OCP International Partnership Association, Inc, http://www.OCPIP.org.

[18] D. A. Mathaikutty and S. K. Shukla, "SoC Design Space Exploration Through Automated IP Selection from SystemC IP Library," *Proceedings of IEEE International SoC Conference (SOCC'06)*, Austin, TX, September 2006, pp. 109–110.

[19] Spirit Consortium, "SPIRIT Schema Working Group," http://www.spiritconsortium.org/.

[20] B. DuCharme, *XML: The Annotated Specification*, Upper Saddle River, NJ: Prentice Hall, 1998.

[21] The GreenSoCs Project, http://www.greensocs.com/.

[22] W. Klingauf, R. Gunzel, O. Bringmann, P. Parfuntseu and M. Burton, "GreenBus — A Generic Interconnect Fabric for Transaction Level Modelling," *Proceedings of 43rd ACM/IEEE Design Automation Conference (DAC'06)*, July 2006, pp. 905–910.

[23] OSCI Group, SystemC Web site, http://www.systemc.org/.

[24] F. Doucet, S. Shukla, M. Otsuka, and R. Gupta, "BALBOA: A Component Based Design Environment for System Models," *IEEE Transactions on Computer Aided Design of Integrated Circuits and Systems*, Vol. 22, No. 12, December 2003, pp. 1597–1612.

[25] F. Doucet, S. Shukla, M. Otsuka, and R. Gupta, "Typing Abstractions and Management in a Component Framework," *Proceedings of Asia and South Pacific Design Automation Conference (ASPDAC'03)*, 2003, pp. 115–122.

[26] FERMAT Group, MCF Web site, http://fermat.ece.vt.edu/MCF/MCF.htm, 2005.

[27] D. Mathaikutty, H. Patel, S. Shukla, and A. Jantsch, "EWD: A Metamodeling Driven Customizable Multi-MoC System Modeling Framework," *ACM Transactions on Design Automation of Electronic Systems*, Vol. 12(3), No. 33, 2007, pp. 1–43.

[28] D. Mathaikutty and S. Shukla, "A Metamodel for Component Composition," FERMAT Lab., Virginia Tech, Tech. Rep. 2005-13, December 2005.

[29] D. Mathaikutty and S. Shukla, "MCF: A Metamodeling based Component Composition Framework — Composing SystemC IPs for Executable System Models," *IEEE Transactions on Very Large Integration (VLSI) Systems*, Vol. 16, No. 7, July 2008, pp. 792–805.

[30] D. Mathaikutty and S. Shukla, "Mining Metadata for Composability of IPs from SystemC IP Library," *Design Automation for Embedded Systems*, Vol. 12, No. 1-2, 2008, pp. 63–94.

[31] Synopsys Inc., "DesignWare® Verification IP Solutions," http://www.synopsys.com/products/designware/vip_solutions.html.

[32] System Verilog, http://www.systemverilog.org/.

[33] Verilog, http://www.verilog.com/.

[34] VHDL, http://www.vhdl.org/.

[35] Synopsys Inc., Open Vera Web site, http://www.open-vera.com/.

[36] R. Goering, "IEEE Catches the Spirit of IP Reuse," *EE Times*, June 2006.

[37] Cadence Inc., "Incisive® Verification IP," http://www.cadence.com/products/.

[38] Cadence Inc., http://www.cadence.com/.

[39] Cadence Inc., "e Verification Language," http://www.cadence.com/partners/industry_initiatives/e_page/index.aspx.

[40] Cadence Inc., Incisive® Enterprise Scenario Builder Web site, http://www.cadence.com/products/functional_ver/scenario_builder/index.aspx.

[41] Verisity Inc., Specman Elite Web site, http://www.verisity.com/products/specman.html.

[42] B. Stohr, M. Simmons, and J. Geishauser, "FlexBench: Reuse of Verification IP to Increase Productivity," *Proceedings of the Design, Automation and Test in Europe Conference and Exhibition (DATE'02)*, 2002.

[43] S. Olcoz, F. Ruiz, A. Gutierrez, and L. Teres, "Design Reuse Means Improving Key Business Aspects," *Proceedings of the VHDL International Users Forum (VIUF) Fall Workshop*, June 1999, pp. 78–84.

[44] P. Rose, "Intellectual Property Reuse – A New Business Model," D & R Industry Articles, http://www.us.design-reuse.com/articles/article5230.html, April 2003.

[45] R. Wilson, "Fabless Semiconductor Companies: Is Intellectual Property Reuse Creating a New Model?" D & R Industry Articles, http://www.us.design-reuse.com/news/news14904.html, December 2006.

[46] M. Horne, "Verification IP Qualification and Usage Methodology for Protocol-Centric SoC Design," D & R Industry Articles, http://www.us.design-reuse.com/articles/article9728.html, February 2005.

[47] Meta-Modeling and Semantic Modeling Community, http://www.metamodel.com.

[48] Wikipedia Foundation, http://en.wikipedia.org/wiki/Meta-modeling.

[49] A. Ledeczi, M. Maroti, A. Bakay, G. Karsai, J. Garrett, C. Thomason, G. Nordstrom, J. Sprinkle, and P. Volgyesi, "The Generic Modeling Environment," *Proceedings of Workshop on Intelligent Signal Processing* Budapest, Hungary, May 2001.

[50] Object Management Group, "Unified Modeling Language," http://www.uml.org/.

[51] J. Warmer and A. Kleppe, *The Object Constraint Language: Getting Your Models Ready for MDA*, 2nd Edition, Reading, MA: Addison-Wesley, 2005.

[52] D. Mathaikutty, S. Kodakara, A. Dingankar, S. Shukla, and D. Lilja, "MMV: Metamodeling Based Microprocessor Validation Environment," *IEEE Transactions on Very Large Integration (VLSI) Systems*, Vol. 16, No. 4, April 2008, pp. 339–352.

[53] D. Mathaikutty and S. Shukla, "On the Composition of IPs for Automated System Model Construction," *Proceedings of Technology and Talent for the 21st Century (TECHCON'07)*, Austin, TX, September 2007.

[54] D. Mathaikutty and S. Shukla, "Type Inference for Component Composition," *Proceedings of 5th ACM-IEEE International Conference on Formal Methods and Models for Codesign (MEMOCODE'07)*, May 2007, pp. 61–70.

[55] D. Mathaikutty and S. Shukla, "A Property-Based Checker Generation Approach for IP Composition," FERMAT Lab., Virginia Tech, Tech. Rep. 2007-14, June 2007.

[56] D. Mathaikutty, A. Dingankar, and S. Shukla, "A Metamodeling Based Framework for Architectural Modeling and Simulator Generation," *Proceedings of Forum on Specification and Design Languages (FDL'07)*, Barcelona, September 2007.

[57] S. Kodakara, D. Mathaikutty, A. Dingankar, S. Shukla, and D. Lilja, "Model Based Test Generation for Microprocessor Architecture Validation," *Proceedings of the 20th International Conference on VLSI Design*, January 2007, pp. 465–472.

[58] D. Mathaikutty, S. Ahuja, A. Dingankar, and S. Shukla, "Model-Driven Test Generation for System Level Validation," *Proceedings of IEEE International Workshop on High Level Design Validation and Test (HLDVT07)*, November 2007, pp. 83–90.

[59] K. Hayhurst, *A Practical Tutorial on Modified Condition/Decision Coverage*, National Aeronautics and Space Administration, Langley Research Center, 2001.

[60] D. Mathaikutty, S. Kodakara, A. Dingankar, S. Shukla, and D. Lilja, "Design Fault Directed Test Generation for Microprocessor Validation," *Proceedings of Design Automation and Test Conference Europe (DATE'07)*, March 2007, pp. 761–766.

[61] S. Kodakara, D. Mathaikutty, A. Dingankar, S. Shukla, and D. Lilja, "A Probabilistic Analysis For Fault Detectability of Code Coverage Metrics," *Proceedings of the 7th International Workshop on Microprocessor Test and Verification (MTV'06)*, December 2006.

[62] M. F. Lau and Y. T. Yu, "An Extended Fault Class Hierarchy for Specification-Based Testing," *ACM Transactions on Software Engineering and Methodology*, Vol. 14, No. 3, 2005, pp. 247–276.

Chapter 2

Background

In this chapter, we introduce the background material necessary to better appreciate the content of this book. We begin with an elaborate discussion on metamodeling that is the key enabling ingredient in our solutions, its classification and a particular academic environment for metamodeling called Generic Modeling Environment (GME). The rest of the chapter sets up the reader with the concepts, techniques, languages, and frameworks that are essential to comprehend the solutions for the problems addressed in Part I and Part II of this book.

2.1 METAMODELING

A *metamodeling framework* facilitates the description of a modeling domain by capturing the domain-specific syntax and semantics into an abstract notation called the *metamodel*. This metamodel is used to create a *modeling language* that the designer can use to construct domain-specific models by conforming to the *metamodeling rules*. These rules are governed by the syntax and "static semantics" of the domain. The framework that enables this is called the *modeling framework*. *Static semantics* [1] are wellformedness conditions specified as invariants that must hold for any model created using the modeling language. The facility of a metamodeling framework and a metamodel put together is called the *metamodeling capability* as in Figure 2.1. We identify two distinct roles played in a metamodeling framework. The role of a *metamodeler* who constructs domain-specific metamodels and the role of a *modeler* who instantiates these metamodels for creating domain-specific models.

Consider a simple audio signal processing example. The metamodel specifies the following types of entities: microphones, preamps, power amps, and loudspeakers. Microphones has one output port, preamps and power amps each have a single input and output port, and a *loudspeaker* has a single input port. The metamodel also specifies the following relationships among the types: a microphone's output port may be connected to the input ports of any number of preamps, a preamp output port may connect to any number of power amp input ports, and a power amp output port may connect to one or more loudspeaker input ports. Such syntactic model construction rules are enforced by the metamodel during modeling and the design editor will not allow any objects or relationships not specified in the metamodel to be part of a model being constructed under that metamodel. Therefore, if a designer modeling an audio system tries to connect a microphone's output to the input port of the loudspeaker (in reality this might lead to damaging the loudspeaker), the metamodel will flag a violation and revert the model back to the state it was before the violation occurred.

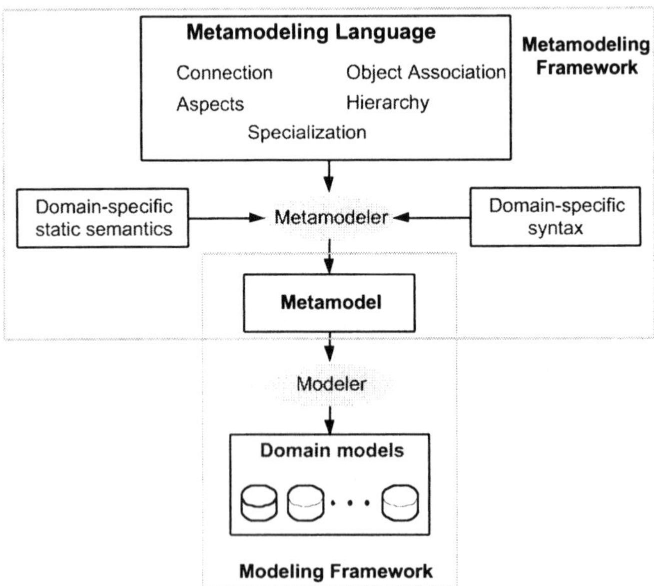

Figure 2.1 Metamodeling capability.

We extend the notion of metamodeling to a design environment, where this capability is used to create and customize metamodels (modeling languages) that are employed in describing systems. Design environments such as Ptolemy II [2], Metropolis [3], and EWD [4] have different degrees of the metamodeling capability; therefore this characteristic of a design environment can be measured based on its ease of customization. An environment that can easily customize metamodels to enhance or restrict the modeling capability is said to be high in its metamodeling capability. An alternate notion of distinction is based on whether the metamodeling is implicitly provided versus it being an explicit capability.

2.1.1 Implicit Metamodeling Versus Explicit Metamodeling

A design environment with *explicit metamodeling* capability must be associated with a metamodeling language, which provides an abstract syntax without implementation details that can be customized into specific metamodels. Such an environment allows one to create metamodels by specifying a collection of modeling object types, along with the relationships allowed between those object types and the attributes associated with the objects. Furthermore, such a framework could enforce the metamodel during the modeling activity through metamodeling rules. Therefore, the modeler has to explicitly state the metamodel employed in the modeling process and any deviation from the metamodel during modeling is notified on the fly and must be corrected.

A design environment has *implicit metamodeling* capability if it does not have a metamodeling language for domain-specific description and makes use of the underlying programming language for the purpose. For example, in Ptolemy II [2], if the modeler wants to customize any existing modeling language by adding newer specialized modeling constructs, then the underlying programming language Java must be employed, because it does not have a metamodeling language of its own.

A summary of the essential concepts of a metamodeling language from [5] are shown in Table 2.1.

The software solutions MCF and MMV described in this book were built on top of an existing academic environment for metamodeling, which provided our work with the necessary metamodeling language and framework.

Table 2.1
Essential Concepts of a Metamodeling Language

Concepts	Description
Connection	Provides rules for connecting objects together and defining interfaces. Used to describe relationships among objects.
Aspects	Enables multiple views of a model. Used to allow models to be constructed and viewed from different perspectives.
Hierarchy	Describes the allowed encapsulation and hierarchical behavior of model objects. Used to represent information hiding.
Object Association	Binary and n-ary associations among modeling objects. Used to constrain the types and multiplicity of connections between objects.
Specialization	Describes inheritance rules. Used to indicate object refinement.

2.1.2 Generic Modeling Environment

The Generic Modeling Environment (GME) [6] is a configurable toolkit that facilitates the easy creation of domain-specific modeling and program synthesis environment. It is proposed as a design environment that is configurable for a wide range of domains and overcomes the high cost associated with developing domain-specific environments, such as MATLAB/Simulink for signal processing and LabView for instrumentation [7]. The metamodeling framework of GME has a *meta-metamodel* that is a set of generic concepts, which are abstract enough to be common to most domains. These concepts can then be customized into a new domain such that they support that domain description directly. The customization is accomplished through metamodels specifying the modeling language of the application domain. It contains all the syntactic, semantic, and presentation information regarding the domain and defines the family of models that can be created using the resultant modeling framework. These models can then be used to generate the applications or to synthesize input to different commercial off-the-shelf analysis tools.

For clarity, we reiterate a few definitions for terms used in the context of GME:

1. A *modeling framework* is a modeling language for creating, analyzing, and translating domain-specific models.
2. A *metamodel* defines the syntax and "static semantics" of a particular application-specific modeling framework.
3. A *metamodeling framework* is used for creating, validating, and translating metamodels.
4. A *meta-metamodel* defines the syntax and semantics of a given metamodeling framework.

The generic concepts describe a system as a graphical, multiaspect, attributed entity-relationship (MAER) diagram. Such a MAER diagram is indifferent to the dynamic semantics of the system, which is determined later during the model interpretation process. The generic concepts supported are *hierarchy*, *multiple aspects*, *sets*, *references*, and *explicit constraints*, which are described using the following constructs.

A *project* (<<Project>>) contains a set of folders that acts as containers helping to organize a collection of objects. The elementary objects are called *atoms* (<<Atom>>) and the compound objects that can have parts and inner structure are called *models* (<<Model>>). These objects are instantiated to be a specific *kind* with a predefined set of *attributes*. The *kind* describes the role played by these objects and the *attributes* are used to annotate them with domain-specific characterizations. The parts of a *model* are objects of type *atom* or other *models*. *Aspects* primarily provide visibility control. Every design has a predefined set of *aspects* describing different levels of abstraction. The existence of a part of the domain within a particular aspect is determined by the metamodel. The simplest way to express the relationship between two objects in GME is with *connections* (<<Connection>>). A connection establishes an annotated relation between a set of objects through predefined attributes. *Constraints* in GME are articulated based on the predicate expression language called OCL [8]. They are used to express relationship restrictions as well as rules for containment hierarchy and the values of the properties ("static semantics" of the domain).

In the following sections, we briefly discuss the background material necessary to understand the problem associated with fast and effective integration of reusable design IPs into a SoC and our solution presented in Part I of this book. Our solution is a component composition framework that helps designers to quickly and correctly create efficient simulation models through automated reuse.

2.2 COMPONENT COMPOSITION FRAMEWORK

A component composition framework (CCF) [9] allows designers to architect the system in terms of components and their composition. Such a framework makes the designer oblivious to the specifics of the software language and allows realizing a system through *design templates* and *tool support*. The design template structurally reflects the modeling abstractions such as register-transfer level (**RTL**), transaction level (**TL**), mixed-abstractions (**RTL-TL**), and hierarchy. The tools facilitate selection and connection of the correct components, automatic creation of interfaces and bridging transactors, simulation of the composed design, and finally testing and verification for correctness of the integration.

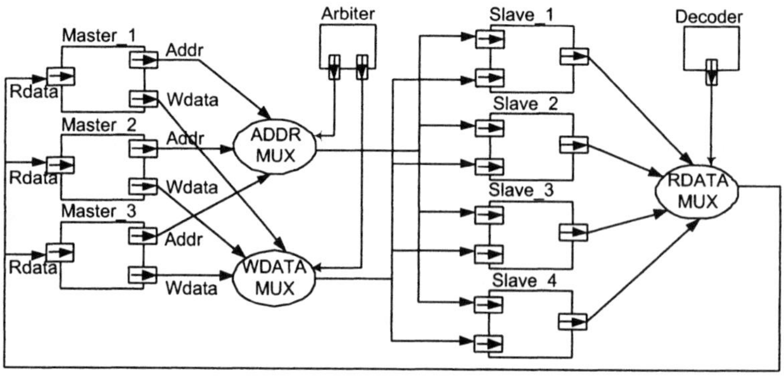

Figure 2.2 RTL AMBA AHB bus model.

In Figures 2.2 and 2.3, we illustrate the use of *design templates* through examples such as the RTL model of the AMBA AHB [10] and TL model of the simple bus from the SystemC [11] distribution. The AMBA bus model shown in Figure 2.2 consists of three masters and four slaves with same interface descriptions that communicate through multiplexed signals selected by an arbiter and a decoder. For creating this model as a description and later instantiating as an executable using reusable components the CCF should allow for the following in its description process: (1) generic RTL components that would be reused for the masters and slaves, (2) generic mux and demux with automatic interface inference for multiplexing and demultiplexing, (3) RTL components that describe the arbiter and decoder, and (4) signal-level connectivity of the components.

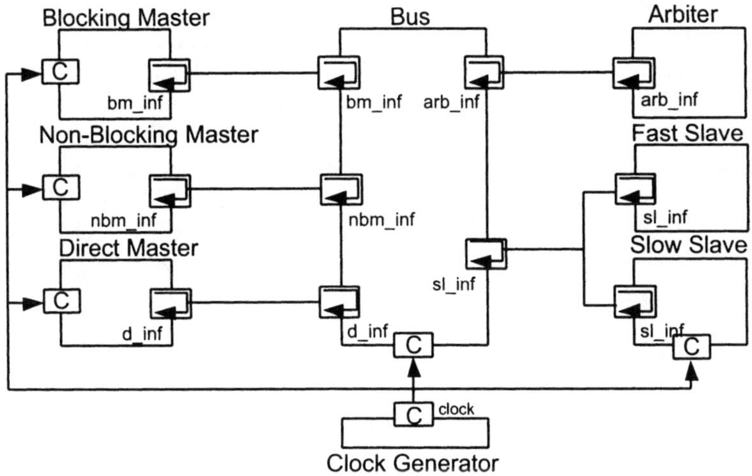

Figure 2.3 TL simple bus model.

Figure 2.3 describes the TL model, which consists of three bus masters with distinct protocol-level interfaces (blocking, nonblocking, and direct read-write access) and an arbiter to schedule their transactions. It consists of two slaves, one with instantaneous read-write access and the other with slow read-write access due to its clock-based activation. To facilitate this, the framework should provide the following in its description process: (1) TL components with function calls that describe the distinct masters, (2) a bus channel that implements the read-write interface of the masters, (3) an arbiter channel that implements the arbitration interface, (4) a channel that implements the read-write interface for the fast slave and which is reused to describe the slow slave, (5) insertion of clock ports for the components and channels, and (6) transaction-level connectivity.

If the designer wants to explore the simple bus model to analyze various trade-offs, then an alternative description could be replacing the slave channels with RTL slaves from the AMBA bus in order to gain more precise timing from the extra handshakes with some loss in simulation efficiency. For such an alternative the framework's description process should allow: (1) modification through reusability of slave components from the AMBA bus, and (2) insertion of transactors that map the TL and RTL interaction of the bus channel and the RTL slaves. Furthermore for the three design scenarios, the framework in general should allow design visualization and enforce checks that look for inconsistencies, datatype mismatches, incompatible interfaces, and illegal component or channel description.

Once such checks are done, the CCF *tool support* should provide automatic selection of IPs at different abstraction levels, synthesize transactors to bridge the abstraction, create testbenches, and integrate all the above to arrive at multiple possible simulation models. The tool support primarily relies on the structural and behavioral characteristics of the reusable design components in the IP library to enable the automated process of selection, configuration, and integration. These characteristics of an IP are called metadata, which are identified and extracted through automated techniques and represented to enable reflection and introspection.

2.3 REFLECTION AND INTROSPECTION (R-I)

Metadata is defined as "data about data" and is the kind of information that describes the characteristics of a program or a model. Information ranging from the structure of a program in terms of the objects contained, their attributes, methods, and properties describing data handling details can be exemplified as metadata. This class of information is necessary for CAD tools to manipulate the intellectual properties (IPs) within system-level design frameworks as well as to facilitate the exchange of IPs between tools and SoC design flows. These exchanges can be efficiently enabled by expressing integration and performance requirements as well as configuration capabilities.

We classify EDA related metadata into two broad categories: (1) interoperability metadata and (2) introspection metadata. Interoperability metadata would enable easier integration of semiconductor IP and IP tools as shown by the consortium named SPIRIT [12], which allows design flow integration by utilizing metadata exchanged in a design-language neutral format. Consider the SPIRIT enabled flow from the ARM RealView SoC Designer tool to coreAssembler, the Synopsys SoC environment in [13]. The transfer is illustrated with an ARM1176JZ-STM processor subsystem design and the metadata captured through SPIRIT include RTL I/O signals, bus-interfaces supported, parametric configurations, abstraction levels, memory map or remap information, interconnect layout, and so forth. This information exported from the SoC Designer allows a seamless import of the ARM1176JZ-STM processor subsystem into the Synopsys coreAssembler. Therefore, interoperability metadata serves as a common standard for exchange of multivendor IPs between tools and design flows.

Introspective metadata on designs in system-level design frameworks allow CAD tools and algorithms to manipulate the designs as well as capture interesting properties about them. These features are also useful for debugging, profiling, type/object browsing, design analyzing, scheme generation, composition validation, type checking, and compatibility checking. *Introspection* is the ability of an executable system to query internal descriptions of itself through some reflective mechanism. The *reflection* mechanism exposes the structural and runtime characteristics of the system and stores it in a data structure. The data stored in this data structure is called the metadata.

For further details on metadata and reflection and introspection techniques, we recommend readers to look at the work of Doucet et al. in [14].

One of the necessary ingredients for a CCF is a library of reusable design components. The IP library employed in MCF is a collection of compiled SystemC designs put together from (1) the SystemC distribution [11] and those publicly available (OPENCORES [15]), (2) our industrial contacts, and (3) the various projects done at the FERMAT Lab [16]. Note that the solution approach presented in this book is not limited to designs specified in SystemC, but can be applied to designs specified in other system-level languages.

2.4 SYSTEMC

SystemC [11] is an open-source modeling and simulation language released by the Open SystemC Initiative (OSCI). SystemC is basically a library of C++ classes that allows modeling at the register transfer level (RTL) and abstractions above RTL. An example of modeling at abstraction levels above RTL is transaction-level modeling (TLM), which is commonly used for fast design exploration and simulation [17]. SystemC's modeling framework consists of SystemC-specific constructs that help in describing the model in a programmatic manner. This program representation of the design is then simulated using SystemC's simulation framework, which follows a discrete event (DE) simulation kernel. A module instantiated using the `SC_MODULE()` macro in SystemC can express a set of SystemC process behaviors. The three types of SystemC processes are `SC_METHOD()`, `SC_THREAD()`, and `SC_CTHREAD()`. These processes are associated with a member function called the *entry* function. When the process triggers, this entry function describes the internals of the process. Each of these processes can be triggered based on their sensitivity list.

For further details in modeling at different levels of abstraction using SystemC, we recommend reading the manuals and documents made available at the SystemC Web site [11].

In the following sections, we discuss the background material necessary to understand Part II, which includes: (1) the problem associated with the reuse of verification IPs to generate collaterals for microprocessor validation, and (2) the solution to enable reuse through generative verification IPs that is different from the traditional validation flow for processors.

2.5 MODEL-DRIVEN VALIDATION

Validation can be performed by using a model-driven approach or a model-checking approach. The model-checking methods [18] take the implementation and create an abstract representation (AR) on which properties are verified using a model-checker. If a property holds in the AR, then it is said to hold in the implementation. Abstraction leads to false counter-examples, even if the abstraction method is sound, and thereby, leading to rounds of abstraction refinement that oftentimes lead to state explosion in the end. Therefore, often an incomplete method such as functional testing is all we can afford to validate a design. On the contrary, the more successful model-driven methods [19, 20] capture the essence of the design as a functional model. This model serves as a formal specification of the desired system (golden reference model) and tests generated from this model are employed in the validation of the system implementation.

2.5.1 Microprocessor Validation Flow

The design teams pass through many levels of abstraction, from system to microarchitecture, when designing a microprocessor. They develop functional specifications for each level of abstraction which then becomes the standard document against which the implementation of the processor is validated. The goal of validation is to ensure that the microcode, RTL, and the fabricated product in silicon implement the behavior defined in the system, architecture, microcode, and microarchitecture specifications [21].

Formal verification and simulation-based validation are the two main techniques used for validating a processor. In formal verification, the specification or implementation is proved to satisfy a given set of properties [22]. Formal verification algorithms suffer from state space explosion problem and hence they cannot handle large designs. When using formal methods, the validator will have to make a trade-off between the level of detail in the model and the capacity of the tool. Formal methods are generally used to validate small but critical hardware blocks in a processor or to validate an abstract version of the implementation [23]. Until formal tools can handle large designs, the simulation-based validation environment will continue to be the workhorse for validating a microprocessor.

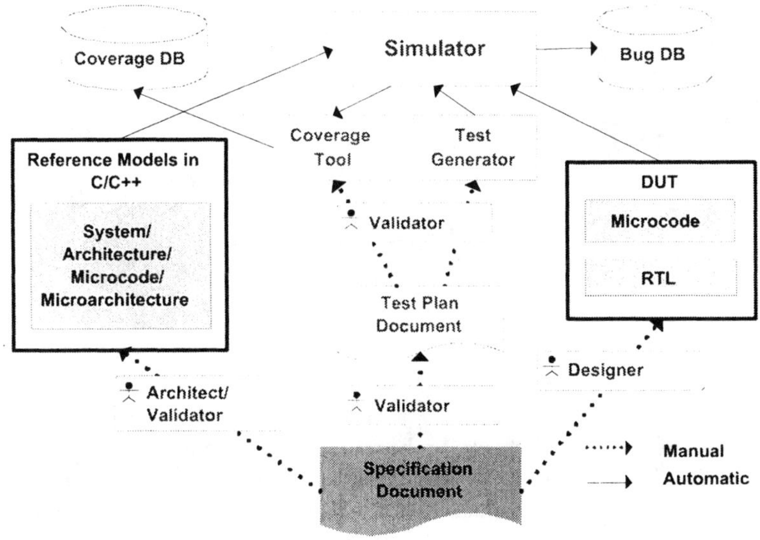

Figure 2.4 Traditional validation environment.

Simulation-based validation is widely used to validate microprocessors in the industry. A traditional simulation-based validation framework is shown in Figure 2.4. In this framework, the models of the microprocessor at different levels of abstraction are manually derived from the specification document. These models act as reference models against which the RTL/microcode/silicon of the processor is validated. A test plan document, which describes the scenarios that need to be tested during validation, is manually derived from the specification. The test plan acts as the reference document for manually guiding the test generators and coverage analysis tools. The generated tests are run on the RTL/microcode/silicon and the validated processor models. The results produced by the test are compared for a mismatch.

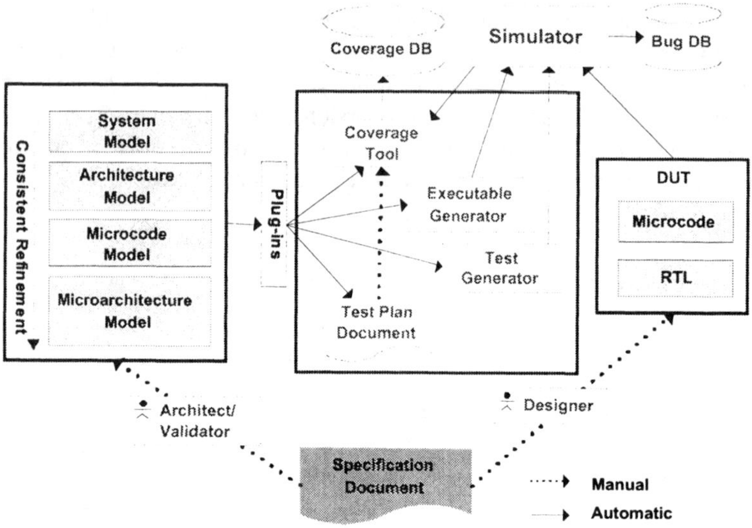

Figure 2.5 Proposed validation environment.

There are several drawbacks in the traditional validation environment shown in Figure 2.4. First, most of the models and tools are generated from the specification document or the test plan. The language in the documents could be ambiguous and could be misinterpreted by validators developing different models and tools. Second, rapid changes in the specification are common during the design and development process. These changes should be manually propagated into the different models to achieve consistency across abstraction levels and tools making it an erroneous process. Third, the models in a traditional validation environment are developed in a modeling framework that depends on the target usage of the model. For example, models used to generate simulators are typically written in programming languages like C/C++ and those for test generation are written in specification languages such as Specman e [24]. Modeling in a specific programming language makes the modeling process highly dependent on the artifacts specific to the underlying programming framework. For example, modeling frameworks based on C/C++ are operational in nature. They do not offer formal semantics for refinement of the models and there is no proof of correctness for the defined refinement maps. In the proposed environment shown in Figure 2.5, *only the models* are derived manually from the specification. All the tools that are used in the validation environment are automatically generated from the models; hence these models are called generative verification IPs.

2.5.2 Simulation-Based Functional Validation

Functional validation is typically done on system, architecture, microcode, and RTL abstractions. In the past, the processor models at the architecture and microcode level were comparatively simpler and easier to design and validate when compared to RTL models. Designing and validating an RTL model is hard due to the presence of microarchitectural performance optimizations like out-of-order processing, super-scalar, and speculative execution and due to the parallel execution model of hardware. Current trends in microprocessor design, like multiple processing cores and/or threads, and support for software applications like virtualization [25] and security [26], increase the design and validation complexity at processor abstraction levels above RTL.

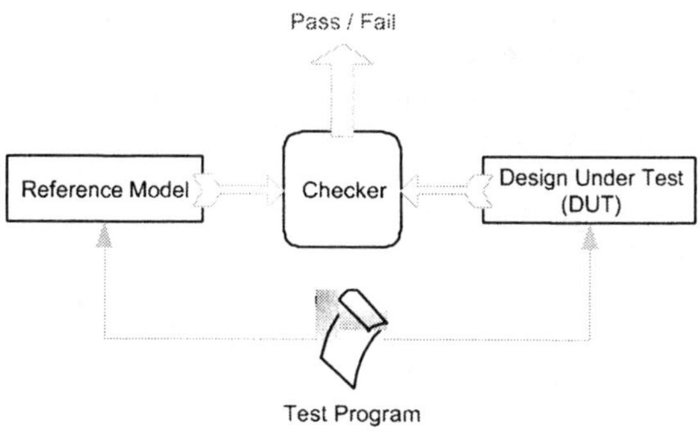

Figure 2.6 Simulation environment.

Simulation is widely used to validate large systems like microprocessors. In simulation-based validation, shown in Figure 2.6, a test is executed against the golden reference model as well as against the design under test (DUT). A bug is recorded if the checker detects a difference in the results between the two. A simulation-based validation environment for microprocessors consists of test generators, simulators for reference models, coverage analysis tools, collection of test cases and test suite, test plans, and so forth.

One of the main verification collaterals generated by our solution is a test generator. We perform different kinds of test generation such as random, structural and functional coverage-directed as well as design fault-directed.

2.6 TEST GENERATION

A test case is defined as a set of inputs, execution conditions, and expected outputs that are developed for a particular objective, such as to exercise a particular path in the implementation or to verify its compliance with a specific requirement. A test suite is a collection of test cases. Test generation for simulation-based validation can be classified into exhaustive, random, manual, and coverage-directed. Exhaustive testing is guaranteed to catch all bugs in the implementation but it is not practical due to its exponential complexity. Random test generators are extensively used to validate microprocessors, but they are often found to generate test cases that exercise the same feature of the processor over and over again. In the manual approach, interesting testing scenarios are captured in a test plan and the test writer manually creates the tests to cover them. This approach is time-consuming, error-prone, and not scalable. Coverage-directed test generation techniques [18, 27, 28] complement the manual approach. These techniques automate the path from coverage to test generation. Test suites that satisfy the coverage metric are automatically generated and they differ in the coverage metric being targeted and the methodology employed to generate test cases.

The presented solution to Part II performs test generation in two different ways: (1) by converting the verification IP into a set of constraints and (2) by specifying the verification IP in the Esterel language [29] and utilizing the esVerify tool of Esterel Studio [30].

2.6.1 Constraint Programming

Constraint programming (CP) is a study of computational systems based on constraints [31]. We model test generation as a constraint satisfaction problem (CSP). The formal definitions of a constraint and CSP are given next.

Constraint and Constraint Satisfaction Problem

A constraint is defined as a logical relation among several variables, each taking a value from a given domain. A constraint is formally defined as follows:

- Let $X = \{x_1, x_2, ...x_n\}$ be a set of variables.
- Each x_i is associated with a finite set of possible values D_i. D_i is also called the *domain* of x_i.
- A constraint C on $x_1, x_2, ...x_n$ is a subset of $D_i \times ... \times D_n$.
- A CSP is a set of constraints restricting the values that the variables in X can simultaneously take.

A *solution* to a *CSP* is an assignment of one value to every variable x_i in X from its domain D_i, such that all constraints are satisfied at once.

Esterel Language

Esterel is a language for modeling synchronous reactive systems [29], especially suited for control-dominated systems. It follows a synchronous programming model [32] with imperative style. It has various constructs to express concurrency, communication, and preemption, whereas data-handling follows the style of procedural languages such as C. Esterel has two different specification styles, namely imperative and equational.

Its semantics can be understood as a Finite State Mealy Machine (FSM), but it must ensure determinism, so that a program generates the same output sequence for the same input sequence. Internally, the Esterel compiler translates the control part of a program into a set of Boolean equations with Boolean registers. For further details on the synchronous formalism and the modeling syntax, we recommend reading the reference manuals and technical documents made available at the Esterel Technologies Web site [33].

2.6.2 Esterel Studio

Esterel Studio (ES) [30] is a development platform for designing reactive systems, which integrates a GUI for design capture, a verification engine for design verification, and code generators to automatically generate target-specific executables. The GUI enables modeling through the graphical state machine specification called SSM or the Esterel textual specification [29]. ES performs static as well as runtime consistency checks on this model for nondeterminism, dead code, and so on. The next most important ingredient of ES is the formal verifier (*esVerify*), which verifies the designer's intent. It allows both assertion-based verification as well as sequential equivalence checking. User-defined properties (expressed as assertions) and automatically extracted properties (out-bound, overflow, and so forth) are formally verified by generating the appropriate counter-traces (.esi files) to illustrate violations. An assertion is flagged as an ***Assume*** to notify ***esVerify*** that it is an assumption about the environment. It can be applied as a constraint during formal verification (FV) of the other non-***Assume*** assertions. The formal methods utilized by *esVerify* are based on bounded, full model checking as well as symbolic model checking. Finally, ES performs multitargeted code generation, which ranges from targets such as RTL (in VHDL/Verilog) to system-level (in C).

2.7 COVERAGE-DIRECTED TEST GENERATION

The effectiveness of the test generation approach is highly dependent on the quality of the test suite used for the validation. Coverage metrics inspired from software validation (statement, branch, and path coverage), state machine representation (state, transition, and path coverage), and functionality of the processor are used to measure the quality of tests. Therefore, to improve the quality of the test suite generated, we also developed a coverage-directed test generator as a verification collateral. The effectiveness of coverage-directed test generation in finding a bug depends the following:

1. Coverage metric chosen;
2. Type of bug/fault present in the design.

2.7.1 Structural Coverage

Structural coverage criteria are used to determine if the code structure is adequately covered by a test suite. Research in software testing has shown that structural coverage criteria is effective in finding large classes of bugs [34]. Since sequential execution of instructions at architecture abstraction is similar to sequential execution of software, we decided to use structural coverage criteria for generating coverage-directed tests. We define three different structural coverage criteria next.

Statement Coverage

Statement coverage is used to verify if every statement in the code is reachable. To achieve statement coverage, a test suite should invoke every executable statement in the code. Statement coverage does not take control flow into consideration and hence is considered as a weak coverage metric.

Branch Coverage

Branch coverage is used to test the control constructs in the code. Branch coverage mandates that the test suite should execute both the *true* and *false* outcomes of all *decisions* (e.g., *if* statements) in the code. This metric is stronger than statement coverage, but is still weak due to the presence of *conditions* within a *decision.* A *condition* is defined as a Boolean expression that contains *no* Boolean operators. For example, $(A > 10)$ is a *condition.* A *decision* could contain zero or more Boolean operators. For example, if $(A > 10||B < 5)$ is a decision. Branch coverage is satisfied in this example if the decision takes both *true* and *false* values. But branch coverage does not take into consideration which *condition* (i.e., $A > 10$ or $B < 5$) within the *decision* caused it to be *true* or *false.*

MCDC

MCDC requires that each *condition* within a *decision* should *independently* affect the outcome of the *decision.* MCDC criteria is a requirement that should be satisfied when verifying critical software used to control avionics [35]. MCDC is formally defined as follows [36].

- Every point in the entry and exit in the program has been invoked at least once.
- Every *condition* in a *decision* in the program has taken all possible outcomes at least once.
- Every *decision* in the program has taken all possible outcomes at least once.
- Each *condition* in a *decision* has been shown to *independently* affect that *decision's* outcome.

The key element in the definition is the last point, which ensures that the effect of each *condition* is tested relative to other *conditions*. In other words, by satisfying MCDC, no *condition* in a *decision* is left *untested* due to logical masking. MCDC is the strongest and most complicated coverage criteria when compared to statement or branch coverage. Statement and branch coverage are automatically covered by satisfying MCDC.

The definition of MCDC states that every *condition* in a *decision* has to be shown to *independently* affect that decision's outcome. In order to test a decision for MCDC we need a pair of test patterns for every condition. One test pattern will set the output of the decision to *false* and the other will make it *true* and the difference between the two test patterns is the state of the *condition* being tested. In this section, we show how fundamental logical operators AND, OR, and NOT are tested for MCDC [36].

- AND operator is sensitive to *false* inputs; that is, if an input is *false*, the output is *false* irrespective of the state of other inputs and the output is *true* only when all inputs are *true*. In order to test an n-input AND operation for MCDC, we would need n test cases that set the output of the AND operation to *false*, and one test case that sets the output to *true*. Each of the n test cases will have one input set to *false* and all other input set to *true*. Thus we need $n + 1$ test cases to test an n-input AND operation for MCDC.
- Testing OR operation for MCDC is similar to testing AND operation. OR operator is sensitive to *true* inputs; that is, if an input is *true*, the output is *true* irrespective of the state of other inputs. The output of an OR operation is *false* only when all inputs are *false*. In order to test an n-input OR operation for MCDC, we would need n test cases that set the output of the OR operation to *true*, and one test case that sets the output to *false*. Each of the n test cases will have one input set to *true* and all other input set to *false*. Thus we need $n + 1$ test cases to test an n-input OR operation for MCDC.
- To test a NOT gate for MCDC, we need two test cases. The input is set *false* with the output observed to be *true* and vice versa.

Consider a decision d with three conditions $(A < 0)$, B and $(C <> 10)$, the test cases generated for this example are shown here:

Example:
$d = ((A<0) \wedge B \vee (C <> 10))$
Test cases generated for MCDC:
case $(A<0)$: $T_1 = [0, 1, 0]$ $(d{=}0)$ and $T_2 = [1, 1, 0]$ $(d{=}1)$
case B: $T_3 = [1, 0, 0]$ $(d{=}0)$ and $T_4 = [1, 1, 0]$ $(d{=}1)$
case $(C <> 10)$: $T_5 = [0, 1, 0]$ $(d{=}0)$ and $T_6 = [0, 1, 1]$ $(d{=}1)$
One possible MCDC test suite: $\{ T_1, T_2, T_3, T_6 \}$

2.7.2 Functional Coverage

In a typical verification flow, code coverage is a means to begin but not end the complete verification; hence it is common to have code coverage complemented by functional coverage. Functional coverage focuses on the functionality of the design and checks if all the important characteristics of the design's functionality have been tested. Industrial efforts such as Specman Elite [24] rely on user-defined events derived from the test plan to estimate functional coverage. In other efforts such as [18], a functional fault model is used to define functional coverage and the test generation is geared towards detecting these design faults. Assertions are used to monitor the behavior of the DUT and to detect interesting legal events for functional coverage [37]. In recent years, assertion-based verification (ABV) techniques that use temporal languages such as Property Specification Language (PSL) [3] have become popular. They allow the user to specify assertions for complex behavioral scenarios that can be seen as potential coverage criteria. Therefore, assertion-based test generation can be used for functional coverage.

2.7.3 Property Specification Language (PSL)

Property Specification Language (PSL) is a language for specifying design properties using concise syntax and well-defined formal semantics [38]. Accellera's PSL (based upon the Sugar language from IBM), is a powerful, concise language for assertion specification and complex modeling. It provides an interoperable specification language to exchange hardware specifications and develop seamless tool integration. It enables the implementor to capture design intent in a verifiable form, while enabling the verification engineer to validate that the implementation satisfies its specification through static or dynamic verification. At the highest level, it is a multipurpose, multilevel, multifavor assertion language. At its lowest level, PSL uses references to signals, variables, and values that exist in the design description. This ensures that each component's full range of behavior will be consistent, and apparent to various industry-standard verification tools, as the component moves through the design chain. In our work, we restrict ourself to the "simple subset" of PSL, which is its linear temporal logic. For further details on the PSL syntax and its various layers, we recommend reading the IEEE Standard 1850 [38].

2.7.4 Fault Classification

We analyzed bug descriptions and modeling errors reported during various studies of microprocessor validation [39,40] to create a good classification. We also related these errors to the fault types observed during software validation by Lau and Yu [41].

They proposed a comprehensive set of nine fault classes related to Boolean expressions in disjunctive normal form (DNF). Let us consider a Boolean expression in DNF form with m terms $S = t_1 + ... + t_m$. Let t_i denote the ith term in S

Table 2.2

Fault Class Descriptions and Expressions

	Fault Class	*Description*	*Expression*
1	Expression Negation Fault (ENF)	A subexpression of S is negated	$S_{ENF}^{t_i+..+t_j} = t_1 + ... + \overline{t_i + .. + t_j} + ... + t_m$, where $1 \leq i \leq j \leq m$
2	Term Omission Fault (TOF)	A term t_i is omitted	$S_{TOF}^{t_i}=t_1+...+t_{i-1}+t_{i+1}+...+t_m$, where $1 \leq i \leq m$.
3	Term Negation Fault (TNF)	A term t_i is erroneously negated	$S_{TNF}^{t_i}=t_1 + .. + \overline{t_i} + ... + t_m$, where $1 \leq i \leq m$
4	Literal Omission Fault (LOF)	A literal x_j^i is omitted from a term t_i	$S_{LOF}^{x_j^i} = t_1 + ... + t_{i-1} + x_1^i...x_{j-1}^i.x_{j+1}^i...x_{k_i}^i + t_{i+1} + ... + t_m$, where $1 \leq i \leq m$ and $1 \leq j \leq k_i$.
5	Literal Insertion Fault (LIF)	An extra literal $x_{k_{i+1}}^i$ is inserted into a term t_i	$S_{LIF}^{x_{k_{i+1}}^i} = t_1+...+t_i.x_{k_{i+1}}^i+...+t_m$, where $1 \leq j \leq k_i$
6	Literal Negation Fault (LNF)	A literal x_j^i is erroneously negated in a term t_i	$S_{LNF}^{x_j^i} = t_1 + ... + t_{i-1} + x_1^i...\overline{x}_j^i...x_{k_i}^i + t_{i+1} + ... + t_m$, where $1 \leq i \leq m$ and $1 \leq j \leq k_i$
7	Literal Reference Fault (LRF)	A literal y is referenced instead of x_j^i in a term t_i	$S_{LRF}^{x_j^i} = t_1 + ... + t_{i-1} + x_1^i...x_{j-1}^i.y.x_{j+1}^i...x_{k_i}^i + t_{i+1} + ... + t_m$, where $1 \leq j \leq k_i$
8	Disjunctive Operator Reference Fault (ORF[+])	A Boolean operator (+) in S is replaced by (.)	$S_{ORF[+]}^{t_i}=t_1+...+t_i.t_{i+1}+...+t_m$, where $1 \leq i \leq m$
9	Conjunctive Operator Reference Fault (ORF[.])	A Boolean operator (.) in S is replaced by (+)	$S_{ORF[.]}^{x_j^i} = t_1 + ... + t_{i-1} + x_1^i...x_j^i + x_{j+1}^i...x_{k_i}^i + t_{i+1} + ... + t_m$, where $1 \leq i \leq m$ and $1 \leq j \leq k_i$

such that t_i is a conjunction of k_i literals. Let x_j^i denote the jth literal in t_i, where $1 \leq j \leq k_i$. S_f^e is defined as the faulty implementation of S, where f is the name of the fault class and e identifies the exact fault in S. Table 2.2 lists the eight different fault classes relevant to our analysis.

References

[1] R. Geisler, "Precise UML Semantics Through Formal Metamodeling," *Proceedings of the Objected Oriented Programming, Systems, Languages, and Applications (OOPSLA'98), Workshop on Formalizing UML. Why? How?*, 1998.

[2] E. A. Lee, "The Ptolemy Project," Web site: http://ptolemy.eecs.berkeley.edu/.

[3] The Metropolis Project Team, "The Metropolis Meta Model Version 0.4," Department of Electrical Engineering and Computer Science, University of California, Berkeley, UCB/ERL M04/38, September 2004.

[4] D. Mathaikutty, H. Patel, S. Shukla, and A. Jantsch, "EWD: A Metamodeling Driven Customizable Multi-MoC System Modeling Framework," *ACM Transactions on Design Automation of Electronic Systems*, Vol. 12(3), No. 33, 2007, pp. 1–43.

[5] G. Nordstrom, J. Sztipanovits, G. Karsai, and A. Ledeczi, "Metamodeling — Rapid Design and Evolution of Domain-Specific Modeling Environment (ECBS'99)," *Proceedings of IEEE Conference and Workshop on Engineering of Computer-Based Systems*, Nashville, TN, April 1999, pp. 68–74.

[6] A. Ledeczi, M. Maroti, A. Bakay, G. Karsai, J. Garrett, C. Thomason, G. Nordstrom, J. Sprinkle, and P. Volgyesi, "The Generic Modeling Environment," *Proceedings of Workshop on Intelligent Signal Processing* Budapest, Hungary, May 2001.

[7] J. Cuny, R. Dunn, S. T. Hackstadt, C. Harrop, H. H. Hersey, A. D. Malony, and D. Toomey, "Building Domain-Specific Environments for Computational Science: A Case Study in Seismic Tomography," *International Journal of Supercomputing Applications and High Performance Computing*, Vol. 11, No. 3, 1997.

[8] J. Warmer and A. Kleppe, *The Object Constraint Language: Getting Your Models Ready for MDA*, 2nd Edition, Reading, MA: Addison-Wesley, 2005.

[9] F. Doucet, S. K. Shukla, M. Otsuka, and R. Gupta, "An Environment for Dynamic Component Composition for Efficient Co-Design," *Proceedings of Design Automation and Test Conference Europe (DATE'02)*, 2002.

[10] ARM, "AMBA Specification 2.0," http://www.arm.com, 1999.

[11] OSCI Group, SystemC Web site, http://www.systemc.org/.

[12] Spirit Consortium, "SPIRIT Schema Working Group," http://www.spiritconsortium.org/.

[13] P. Grun, C. Baxter, M. Noll, and G. Madl, "Integrating a Multi-Vendor ESL-to-Silicon Design Flow Using SPIRIT," D & R Industry Articles, http://www.us.design-reuse.com/articles/article12331.html, January 2006.

[14] F. Doucet, S. Shukla, and R. Gupta, "Introspection in System-Level Language Frameworks: Meta-Level Vs. Integrated," *Proceedings of Design, Automation and Test in Europe Conference and Exhibition (DATE'03)*, 2003, pp. 382–387.

[15] OPENCORES Initiative, www.opencores.org.

[16] FERMAT Group, http://fermat.ece.vt.edu.

[17] F. Ghenassia, *Transaction-Level Modeling with SystemC: TLM Concepts and Applications for Embedded Systems*, 1st Edition, New York: Springer-Verlag, 2006.

[18] P. Mishra and N. Dutt, "Functional Coverage Driven Test Generation for Validation of Pipelined Processors," *Proceedings of the Conference on Design, Automation and Test in Europe (DATE'05)*, 2005, pp. 678–683.

[19] H. Patel and S. Shukla, "Model-Driven Validation of SystemC Designs," *Proceedings of the Design Automation Conference (DAC'07)*, 2007, pp. 29–34.

[20] A. Habibi and S. Tahar, "Design for Verification of SystemC Transaction Level Models," *IEEE Transactions on Very Large Scale Integration Systems*, Vol. 14, No. 1, 2006, pp. 57–68.

[21] B. Bentley, "Validating the INTEL Pentium® 4 Microprocessor," *Proceedings of the 38th Conference on Design Automation (DAC'01)*, 2001, pp. 244–248.

[22] E. M. Clarke, O. Grumberg, and D. A. Peled, *Model Checking*, Cambridge, MA: MIT Press, 1999.

[23] D. Moundanos, J. A. Abraham, and Y. V. Hoskote, "Abstraction Techniques for Validation Coverage Analysis and Test Generation," *IEEE Transactions on Computers*, Vol. 47, No. 1, 1998, pp. 2–14.

[24] S. Iman and S. Joshi, *The e-Hardware Verification Language (Information Technology: Transmission, Processing & Storage)*, Norwell, MA: Kluwer Academic Publishers, 2004.

[25] R. Uhlig, G. Neiger, D. Rodgers, A. L. Santoni, F. C. Martins, A. V. Anderson, S. M. Bennett, A. Kagi, F. H. Leung, and L. Smith, "INTEL® Virtualization Technology," *Computer*, Vol. 38, May 2005, pp. 48–56.

[26] "LaGrande Technology Preliminary Architecture Specification," Intel®, 2006.

[27] M. Benjamin, D. Geist, A. Hartman, Y. Wolfsthal, G. Mas, and R. Smeets, "A Study in Coverage Driven Test Generation," *Proceedings of the 36th Design Automation Conference (DAC'99)*, June 1999, pp. 970–975.

[28] S. Ur and Y. Yadin, "Micro Architecture Coverage Directed Generation of Test Programs," *Proceedings of the 36th ACM/IEEE Conference on Design Automation (DAC'99)*, 1999, pp. 175–180.

[29] G. Berry, "The Foundations of ESTEREL," *Proof, Language and Interaction: Essays in Honour of Robin Milner*, 2000, pp. 425–454.

[30] Esterel Technologies Inc., http://www.esterel-technologies.com/products/esterel-studio/.

[31] R. Bartk, "Constraint Programming: In Pursuit of the Holy Grail," *Proceedings of the Week of Doctoral Students (WDS'99)*, MatFyzPress, 1999, pp. 555–564.

[32] G. Berry and G. Gonthier, "The ESTEREL Synchronous Programming Language: Design, Semantics, Implementation," *Science of Computer Programming*, Vol. 19, No. 2, 1992, pp. 87–152.

[33] Esterel Technologies Inc., "Esterel," www.esterel-technologies.com.

[34] A. Dupuy and N. Leveson, "An Empirical Evaluation of the MC/DC Coverage Criterion on the HETE-2 Satellite Software," *Proceedings of the Digital Aviation Systems Conference (DASC'02)*, October 2000.

[35] J. Offutt, Y. Xiong, and S. Liu, "Criteria for Generating Specification-Based Tests," *Proceedings of the 5th IEEE International Conference on Engineering of Complex Computer Systems (ICECCS'99)*, October 1999, pp. 119–129.

[36] K. Hayhurst, *A Practical Tutorial on Modified Condition/Decision Coverage*, National Aeronautics and Space Administration, Langley Research Center, 2001.

[37] A. Ziv, "Cross-Product Functional Coverage Measurement with Temporal Properties-Based Assertions," *Proceedings of the Conference on Design, Automation and Test in Europe (DATE'03)*, 2003, pp. 834–839.

[38] IEEE Std 1850, "IEEE Standard for Property Specification Language (PSL)," IEEE, 2005.

[39] D. V. Campenhout, T. Mudge, and J. Hayes, "Collection and Analysis of Microprocessor Design Errors," *IEEE Design and Test of Computers, IEEE Computer Society Press*, Vol. 17, No. 4, 2000, pp. 51–60.

[40] M. Velev, "Collection of High-Level Microprocessor Bugs from Formal Verification of Pipelined and Superscalar Designs," *Proceedings of the International Test Conference (ITC'03)*, 2003, pp. 138–147.

[41] M. F. Lau and Y. T. Yu, "An Extended Fault Class Hierarchy for Specification-Based Testing," *ACM Transactions on Software Engineering and Methodology*, Vol. 14, No. 3, 2005, pp. 247–276.

Chapter 3

Related Work

This chapter briefly discusses different projects that are related to the solutions presented for the problems addressed in Part I and Part II of this book.

3.1 COMPONENT COMPOSITION FRAMEWORK

There are several projects that employ the idea of component-based design for describing system behavior. Examples of such frameworks are BALBOA [1], LSE [2], Ptolemy II [3], Metropolis [4], and EWD [5]. We briefly describe some of these design frameworks and discuss how our Part I solution differs from these design frameworks.

3.1.1 The BALBOA Framework

The BALBOA framework [1, 6] provides a CCF for building architectural design models. Components are connected and their interface specified via the component integration language (CIL). Before beginning simulation, the framework performs type inference to select appropriate implementation based on the port-connectivity. The CIL is a script-like language used to manipulate and assemble C++ components with constructs that provide an abstraction over C++. Our component composition language (CCL) uses UML class diagrams for describing modeling constructs and OCL to enforce the semantics. Our checkers perform different kinds of static analysis and inference to remove inconsistencies. As a result, our CCL is far more expressive than the CIL. Furthermore, we envision MCF to select components based on behavioral types of the components in the system. Furthermore, our CCL provides design templates for various abstraction levels as opposed to BALBOA, which is meant for RTL.

BALBOA uses their BALBOA Interface Description Language (BIDL) to describe components, very similar to CORBA IDLs. Originally IDLs provided the system with type information, but BALBOA extends this further by providing structural information about the component such as ports, port sizes, and number of processes. This information is stored at a metalayer (a data structure representing the reflected characteristics). The reflection mechanism through the BIDL description is both inconvenient and time-consuming. In our framework, we extract the structural information through automated reflection in a unified manner by preprocessing SystemC models.

BALBOA shows that their type inference problem [7] is NP-complete and propose a heuristic. BALBOA does not allow for polymorphic IPs and their specification enables flexible components only through component overloading. MCF allows both and therefore addresses a larger type inference problem. BALBOA performs type inference by propagating the type information from object to object, which are derived based on the overall system architecture and component connectivity. The type inference problem in MCF needs to be addressed differently because of the methodology. MCF ties in a metamodeling framework that enables distinguishing between entities. Therefore, the compositional properties of these entities are known and taken into account during type inference, which BALBOA is unable to do. MCF also performs type inference based on the system architecture and component connectivity.

3.1.2 Liberty Simulation Environment (LSE)

The Liberty Simulation Environment [2] is a framework for construction of microarchitectural simulation models. A model is structural and concurrent consisting of a netlist of connected instances of components, which communicate through ports. The Liberty framework has a language similar to our CCL, but is devoid of any formalism and the main distinction being that our metamodel driven formalism is rigorous and extensible. Our modeling language is for architectural specification and does not associate any simulation semantics with the components. Furthermore, they do not allow different modeling abstraction besides RTL.

They formulated a type-system [8] for hardware models and propose a heuristic for the type inference problem in their type-system that supports polymorphic and component overloading. The type inference problem is very different, since they want to provide component-based reuse techniques for hardware modeling.

However in LSE the component library is written in a specific manner and hence does not address the IP composition problem given a library of arbitrary SystemC IPs.

Design environments such Ptolemy II [3], Metropolis [4], and EWD [5] describe component behavior using models of computation (MoCs) and have different degrees of metamodeling capability. Therefore, we discuss how our Part I solution differs from these design environments in these aspects.

3.1.3 EWD

EWD [5] is a modeling and simulation framework with support for multiple models of computation (MoCs) [9] and ensures conformance of the MoC formalisms during model construction using a metamodeling approach. In addition, EWD provides a suite of translation tools that generate executable models for SML-Sys [10–12] and ForSyDe [13] simulation framework. The EWD methodology uses GME for customization of the MoC-specific modeling syntax into a visual representation. To embed the execution semantics of the MoCs into the models, they have built parsing and translation tools that leverage an XML-based interoperability language. This interoperability language is then translated into executable Standard ML or Haskell models that can also be analyzed by existing simulation frameworks such as SML-Sys or ForSyDe. In summary, EWD is a metamodeling driven multitarget design environment with multi-MoC modeling capability. EWD has explicit metamodeling capability, since it is built using GME's metamodeling framework. MCF uses GME for defining the visual modeling syntax and semantics for component composition. In order to enable IP selection and composition from the visual model, XML translation tools are built that allow the downstream tools to analyze the component composition model. EWD has explicit metamodeling capability similar to MCF, since it is built using GME's metamodeling framework.

3.1.4 Ptolemy II

Ptolemy II supports multiple modeling domains that facilitate designers to model in a truly heterogeneous manner, therefore it is a multi-MoC modeling framework. Every behavior in Ptolemy II is realized by an MoC [3] and a combination of MoCs can be used to describe a system. Ptolemy II's multi-MoC framework follows an actor-oriented modeling approach, which comes with a Java-based graphical user interface (GUI) through which designers can "drag-and-drop" actors to construct models. Ptolemy II's action-oriented approach [14] has a notion of atomic and composite actors. The atomic actors describe an atomic unit of computation and the composite actors describe a medium through which more complex computation described by other atomic and composite actors can be hierarchically modeled. In Ptolemy II, *directors* implement the MoC behavior and simulate the model. A model or composite actor that follows a particular director is said to be a part of that MoC's domain. Ptolemy has implicit metamodeling capability and the designer is burdened with having to know the underlying programming language (Java) in order to customize the modeling framework.

Ptolemy II follows an inequality formulation, because the type conversion performed using their type-system [15] naturally maps to inequality relations among types. In a component composition framework like MCF, where components are specified at the architectural level devoid of an implementation detail, the framework does not have enough information about the component's behavior, so the components must provide their type information as either type-bindings to their ports, or by mapping undeclared ports to UN, which implies that the type is unknown. Similar to MCF, Ptolemy II requires its actors to either provide the type information on their declared ports or type constraints on their undeclared ports. However during type inference, MCF begins with UN on the undeclared ports of components in the architectural template and automatically creates type-binding on the resolved ports and type constraints on the unresolved ones. This type inference is achieved using information of architectural connectivity as well as the static properties of the instantiated communication primitives (bus, switch, and so forth) in the template. The advantage is a rapid specification process allowing the modeler to create partially specified architectural templates, thereby shifting the complexity to the framework in inferring and constraining the template for IP composition. A comparison with the type inference problem that exists in Ptolemy II is provided in Chapter 6.

3.1.5 Metropolis

The Metropolis metamodel implemented in Java provides computation, communication, and synchronization primitives as abstract classes. Some of these primitives are process, netlist, media, statemedia, and quantity manager. MCF's meta-metamodel concepts are expressed as UML class diagrams, which support techniques such as hierarchy, multiple aspects, sets, references, and explicit constraints. The Metropolis metamodel [4] is independent of any MoC semantics. A design is directly described using this metamodel, but a set of MoC-specific platforms is provided to facilitate the system description. MCF's meta-metamodel is only for customization and the design descriptions are done through metamodels.

Metropolis metamodeling capability is provided by inheritance-based platform derivations, whereas MCF facilitates metamodeling through containment and constraint-based hierarchical decomposition. One of the main advantages of MCF's metamodeling capability is its visualization that we believe eases model construction. A point to note is that MCF enforces metamodeling rules (through its metamodel) that the modeler has to strictly follow during modeling. These "static semantics" [16] cannot be enforced using MoC platforms in Metropolis.

There are several projects that employ the idea of component-based design for describing software systems. We briefly discuss a few of these design environments that enable component reuse for component-based software development and compare our Part I solution with them.

3.2 COMPONENT-BASED SOFTWARE DESIGN ENVIRONMENTS

Component-based software engineering (CBSE) [17, 18] is an approach to constructing software programs using a library of components. The idea is to build programs by composing various library components, rather than building the program from scratch. Since CBSE encourages software reuse, there are potential savings on development time and costs.

SPARTACAS [19] is a framework for specification-based component retrieval and adaptation of software for embedded and digital signal processing systems. The objective of SPARTACAS is to retrieve possible solutions to a problem from a library of components. The behavior of the components in the library is specified without stating the implementation details. The component and problem specifications are done in the same language Rosetta [20]. In this language, a component is described by defining the domain and range of the input and output variables of the component. The operations are described by defining pre- and postconditions on the variables. Being able to retrieve components that match a problem implies that the implementation of the component can be reused to implement the problem. If a component that partially satisfies the requirements of a problem is retrieved, then the component is adapted before reuse. MCF differs from SPARTACAS in many respects. First, the IP library consists of SystemC-based IPs, from which the metainformation necessary for IP composition will be automatically synthesized into XML schemas. The problem specification is in the form of an architectural template, which is described in a component composition language with design templates and multiple abstractions that are geared towards SystemC-based IP integration. Furthermore, the component composition language allows for an architectural description of components and their interactions, which does not include any behavior annotations. The objective of MCF is to solve a problem in the form of an architectural template given a library of SystemC IPs or predict the impossibility of the task. In SPARTACAS, the problem and components in the library are specified in the same language. In MCF, the architectural template needs to be converted to an intermediate representation (XML) so as to allow for IP matching based on extracted metainformation. This is necessary to enable a design flow that ties an arbitrary SystemC IP library.

ABRIE [21] provides a visual environment to describe architectures using components and connectors, which contain a description of its properties. ABRIE uses a semiautomated capability to submit the architecture elements to a reuse engine. Existing components that match the properties of the elements can be reused and instantiated in the architecture. MCF employs a metamodeling paradigm to define its component composition language with architectural elements as well as different hardware abstractions. ABRIE's specification capability is more tiered toward software reuse and uses the model proposed by Shaw et al. [22] for architectural description. MCF automates the construction and instantiation of architectures with matching implementations from a SystemC IP library to satisfy a design problem.

Another approach that is architecture-based similar to ABRIE is discussed in [23]. However, their architectural specification is based on CORBA IDL, which they extend to describe both the services provided and required by a component and their connection mechanism is based on the binding of services. BALBOA [1] employs the same approach, where they use COBRA IDLs to describe their IP implementations.

3.3 IP INTERFACING STANDARDS

A lot of industrial effort has gone into defining IP interfacing standards such as VSIA [24], OCP [25], and SPIRIT [26] to enable design reuse. VSIA [24] specifies interface standards for software and hardware virtual components for the development of SoCs. OCP [25] provides a socket standard with communication socket for an IP to specify: data flow signaling, control signals, and test signals. We briefly compare SPIRIT that provides a standard for IP metadata exchange between tools and design flow with our Part I solution.

3.3.1 SPIRIT

SPIRIT [26] enables system design using a generic metadata model [27] that captures configuration and integration related specifics (*interoperability metadata*) of an IP into an XML library. This enables an SoC design flow in which the modeler instantiates and characterizes these IPs to create the system design. The modeler is required to perform the tedious characterization process, which could be automated by a CCF through selection and composition techniques. As a part of the SoC design flow, SPIRIT provides API standards that should be implemented to export the design and configure the actual IPs to create the executable. Therefore they do not have a tool instead provide the standards for the metadata capture and the configuration and integration APIs. SPIRIT does not have a type inference problem because they depend on user-level characterizations. However, an interface-level type compatible problem does exist, which they handle in their design flow. For MCF the type compatibility problem is a subproblem of the type inference problem.

Our Part I solution primarily relies on the structural metadata of the reusable design components to enable the automated process of selection, configuration, and integration. These characteristics are identified and extracted through structural reflection tools.

3.4 EXISTING TOOLS FOR STRUCTURAL REFLECTION

Several front-end parser suites may be used for implementing structural reflection in SystemC such as EDG [28], Pinapa [29], and KarSCPar [30]. Some structural reflection tools for SystemC also exist such as SystemPerl [31], SystemCXML [32, 33], and ESys.NET [34]. We discuss and compare these approaches below.

EDG [28] is a commercial front-end parser for C/C++ that extracts the C/C++ constructs and populates a data structure. It is further analyzed to interpret SystemC constructs. However, interpretation of SystemC constructs is a complex and time-consuming task, plus EDG is not freely available in the public domain.

Pinapa [29] is SystemC front end that uses GCC's front end to parse all C++ constructs and infers the structural information of the SystemC model by adding a hook to SystemC's elaboration phase. This idea of executing the elaboration phase in order to extract information about the specific design is a very attractive solution. SystemC's elaboration constructs all the necessary objects and performs the bindings after which a regular SystemC model begins simulation. Pinapa examines the data structures of SystemC's scheduler and creates its own internal representation. This is a very good solution for tackling the SystemC parsing issue. However, Pinapa employs an intrusive approach that requires modifications of the GCC source code that make it dependent on changes in the GCC code-base and forbids the use of any other compiler with the benefit of parsing SystemC. In addition, a small change to the SystemC source is also required and they do not have an XML backend.

KaSCPar [30] is a SystemC parser that consists of two components for generating either the AST or a description of the SystemC design, both in XML. The output of the first tool (SC2AST) is an AST that contains a token for each C/C++ and SystemC construct in XML. These tokens capture metainformation about the construct into attributes of the token. These tokens are generated by parsing the output of the preprocessor of GNU GCC. The second tool (SC2XML) takes the AST as input and generates the XML description of the SystemC design. The XML representation consists of the functional and structural characteristics of the analyzed system. The functional part contains information about threads, module constructors, and the control/data flow of each thread. The structural part contains information about the system structure such as top-module, module, port declaration, and connections. The structural types of components, channels, and transactors, as well as the interaction styles are not extracted by SC2XML. It separates the XMLized output of a module from the SC2AST into its structural part (port and hierarchy) and its behavioral part (thread model/method invocations). They do not perform data mining to simplify the typing requirements of the different components in an IP. Therefore, SC2XML was not sufficient for the requirements of a CCF. Furthermore during our initial experimentation with SC2XML, we found the tool to be unstable, even on the SystemC distribution examples. Therefore, we employed SC2AST, which was a stable tool to transform the SystemC/C++ code into XML tags.

Veripool's SystemPerl [31] implements a SystemC parser and netlist generator using Perl scripts. It employs the power of regular expressions for the task of recognizing SystemC constructs. However, one major distinction between EDG and SystemPerl is that SystemPerl only extracts structural information and not behavioral. Furthermore, SystemPerl has some limitations such as it requires source-level hints in the model for the extraction of necessary information and the internal representation of SystemPerl cannot be easily adapted to other environments and purposes.

SystemCXML [32,33] uses an XML-based [35] approach to extract structural information from SystemC models, which can be easily exploited by backend passes for analysis, visualization, and other structural analysis purposes. It uses the documentation system Doxygen [36] and an Open Source XML parser [37] to achieve the structural extraction. The extraction process is very tedious and error-prone, since the documentation for the SystemC IP created through Doxygen is not uniform.

ESys.NET [34, 38] is a system-level modeling and simulation environment using the .NET framework [39] and C# language [40]. There are obvious advantages in making ESys.NET a complete mixed-language modeling framework interoperable with system-level design languages such as SystemC. Unfortunately, one of the major drawback as mentioned earlier is that C# and .NET framework are proprietary technologies of Microsoft. Even though there are open-source attempts at imitating C#, the .NET framework as a whole may be difficult to conceive in the near future. The authors of [34], inspired by the .NET framework's reflection mechanism propose the idea of a composite design pattern for unification of data-types for SystemC. They enhance SystemC's data-type library by implementing the design pattern with additional C++ classes. This altered data-type library introduces member functions that provide introspection capabilities for the particular datatypes. However, this requires altering the SystemC data-type library and altering the original source code to extract structural information. This raises issues with maintainability with version changes, updates, and standard changes due to the highly coupled solution for introspection.

Different simulation-based modeling and validation environments were proposed in the literature. They fall under the category of architecture description languages and frameworks and test generation frameworks. In the following sections, we discuss the related work done on these categories and compare them with our solution in Part II that addresses the problem associated with the reuse of verification IPs to generate collaterals for microprocessor validation.

3.5 ARCHITECTURE DESCRIPTION LANGUAGES

Architecture description languages (ADLs) are narrow spectrum languages. They are designed for a specific target application such as generating an ISA simulator, cycle accurate microarchitecture simulator, and instruction descriptions for the backend of the compiler. Since ADLs are developed for specific applications, it would be difficult to retarget the model developed in an ADL to another application. For example, depending on the ADL, there might not be a neat and consistent way to convert a model written for generating a simulator into constraints that could be used for test generation. Also, these languages do not distinguish between different abstraction levels. This makes it difficult to verify if two models defined in different abstraction levels are consistent with one another.

An extensive body of recent work addresses architectural description language (ADL) driven software toolkit generation and DSE for processor-based embedded systems [41]. These are classified into two categories depending on whether they capture the behavior (Instruction-Set) or the structure of the processor.

nML [42] and ISDL [43] are examples of behavior-centric ADLs. In nML [42], the processor's instruction set is described as an attributed grammar-based derivation. ISDL [43] is similar to nML at the structural level with the capability for explicit constraint modeling to restrict parallelism. Expression [44] is a mixed-level approach that captures the structure, behavior, and mapping (between structure and behavior) of the architecture. At the behavioral level, it allows specification of instructions and operations as well as the operation mapping and at the structural level they describe architecture components, pipeline/data-transfer path, and memory subsystem. LISA [45] is one such ADL whose main characteristic is the operation-level description of the pipeline.

Our modeling framework in MMV facilitates processor modeling through an abstract representation called the *metamodel*. The metamodel defines the syntax and semantic for the language in which the modeler describes the architecture and microarchitecture of the processor. For architectural modeling, the processor is specified through a register/memory-level description (registers, their whole-part relationship, and memory schematic) and an instruction set capture (instruction-behavior).

For microarchitecture modeling, the language allows the modeler to instantiate and configure different pipeline stages, provide ports to interface with the memory and register file, insert instruction registers and describe the control logic needed to perform data forwarding and stalling.

This metamodeling-driven modeling approach enables customizability and uniform analysis for model interpretation. We analyze the model to generate validation collaterals such as simulators, test generators, and coverage tools.

3.6 TEST GENERATION

Test generation for simulation based validation can be broadly classified into a model checker-based approach versus a constraint solver-based approach. In this section, we outline some of the random and coverage-directed test generation performed based on both these approaches.

Adir et al. developed GENESYS-PRO [46] a random test generator, which relies on a model-based approach and test templates coupled with a constraint solver to generate tests. It requires a high level of expertise to model architectures and testing knowledge to use the full power of test templates during test generation. Furthermore, these test templates are mapped from interesting verification scenarios and are not meant for exhaustive testing.

Mishra et al. proposed a model checking-based approach to automatically generate functional test programs for pipelined processors in [47]. The processor model is extracted from the ADL specification and specific properties are applied to the model using an SMV model checker to generate test programs. The model checker is a formal verification tool, it is not clear how the test generator will scale with the size of the design.

Several techniques were proposed in the literature for coverage directed test generation, which differ in the coverage metric being targeted and the techniques used to generate test cases. Ur and Yadin [48] presented a method for coverage-directed generation of assembler test programs that systematically probe the micro architecture of a PowerPC superscalar processor. Benjamin et al. [49] described a verification methodology that uses coverage of formal models to specify tests for a superscalar processor. In [50], Mishra et al. proposed graph-based test generation for validation of pipelined processors that achieves functional coverage.

Our approach [51] proposes a model-based test generation for microcode and RTL validation. The framework allows modeling the processor at the architectural and microarchitectural abstraction and translates these models into constraint satisfaction problems that are resolved through a constraint solver to generate random, coverage-directed, and design fault-directed test suites that validate the microcode and RTL.

Our test generation framework converts the architectural and microarchitectural model into constraint satisfaction problems (CSPs) [52] and passes them through a constraint solver and a test case generator to create test suites. A model created using our modeling framework is converted into a program flow graph, where all the architectural/microarchitectural constructs are mapped to a bunch of states, decisions, and transitions. The generated graph undergoes static single-assignment (SSA) analysis [53] to create the correct set of constraints for the CSP formulation, which is given as input to the constraint solver in the test case generator (TCG). The solution generated by the solver consists of a possible value for every constraint variable in the problem such that the CSP is satisfied. This solution is called a *test case*. The framework also provides a coverage-directed approach that generates one possible test suite, which attains 100% coverage of a specific goal.

For this approach, coverage constraints are given as input to the solver that direct the test generation towards the goal.

In [54, 55], model-driven validation begins by describing the SystemC implementation using the abstract state machine language (AsmL), which requires the discrete-event (DE) semantics to be provided through ASMs. In [54], a designer can visually explore the actions of interest in the ASM model using SpecExplorer and generate tests. These tests are used to drive the SystemC implementation from the ASM model. The test generation capability is limited and not scalable. Our methodology performs functional modeling in Esterel that inherently provides synchronous semantics, which is sufficient for hardware development. We perform coverage-directed test generation that attains structural as well as functional coverage of the implementation. Furthermore, our validation does not require the implementation to be tied to the functional model, which alleviates the tediousness of exporting the implementation APIs into the functional model as in [54]. Furthermore, our TestSpec generation is based on an ATF and SIF, which allows our methodology to be generic and extended for system-level test generation in other validation targets. Our counter example (test) generation performs a reachability analysis; hence it is more scalable with respect to [54] that provides all possible solutions and requires pruning technique for scalability.

References

[1] F. Doucet, S. Shukla, M. Otsuka, and R. Gupta, "BALBOA: A Component Based Design Environment for System Models," *IEEE Transactions on Computer Aided Design of Integrated Circuits and Systems*, Vol. 22, No. 12, December 2003, pp. 1597–1612.

[2] M. Vachharajani, N. Vachharajani, D. Penry, J. Blome, and D. August, "Microarchitectural Exploration with Liberty," *Proceedings of the 35th International Symposium on Microarchitecture (MICRO'02)*, November, 2002.

[3] C. Brooks, E. A. Lee, X. Liu, S. Neuendorffer, Y. Zhao, and H. Zheng, "Heterogeneous Concurrent Modeling and Design in Java," Memorandum UCB/ERL M05/21, University of California, Berkeley, Tech. Rep., July 2005.

[4] The Metropolis Project Team, "The Metropolis Meta Model Version 0.4," Department of Electrical Engineering and Computer Science, University of California, Berkeley, UCB/ERL M04/38, September 2004.

[5] D. Mathaikutty, H. Patel, S. Shukla, and A. Jantsch, "EWD: A Metamodeling Driven Customizable Multi-MoC System Modeling Framework," *ACM Transactions on Design Automation of Electronic Systems*, Vol. 12(3), No. 33, 2007, pp. 1–43.

[6] F. Doucet, S. Shukla, M. Otsuka, and R. Gupta, "An Environment for Dynamic Component Composition for Efficient Co-Design," *Proceedings of Design Automation and Test Conference Europe (DATE'02)*, 2002.

[7] F. Doucet, S. Shukla, M. Otsuka, and R. Gupta, "Typing Abstractions and Management in a Component Framework," *Proceedings of Asia and South Pacific Design Automation Conference (ASPDAC'03)*, 2003, pp. 115–122.

[8] M. Vachharajani, N. Vachharajani, S. Malik, and D. August, "Facilitating Reuse in Hardware Models with Enhanced Type Inference," *Proceedings of the IEEE/ACM/IFIP 2nd International Conference on Hardware/Software Codesign and System Synthesis (CODES+ISSS'04)*, September 2004, pp. 86–91.

[9] A. Jantsch, *Modeling Embedded Systems and SOC's Concurrency and Time in Models of Computation*, Norwell, MA: Morgan Kaufmann Publishers, 2003.

[10] FERMAT Group, SML-Sys Web site, http://fermat.ece.vt.edu/SMLFramework, 2004.

[11] D. Mathaikutty, H. Patel, S. Shukla, and A. Jantsch, "SML-Sys: A Functional Framework with Multiple Models of Computation for Modeling Heterogeneous System," *Design Automation for Embedded Systems*, Vol. 12, No. 1-2, 2008, pp 1–30.

[12] D. A. Mathaikutty, "Functional Programming and Metamodeling Frameworks for System Design," Virginia Polytechnic Institute and State University, Computer Engineering, M.S. Thesis, 2005.

[13] I. Sander and A. Jantsch, "System Modeling and Transformational Design Refinement in ForSyDe," *IEEE Transactions on Computer-Aided Design of Integrated Circuits and Systems*, Vol. 23, No. 1, January 2004, pp. 17–32.

[14] S. Neuendorffer, "Actor-Oriented Metaprogramming," Technical Memorandum UCB/ERL M05/1, University of California, Berkeley, Tech. Rep., January 2005.

[15] Y. Xiong, "An Extensible Type System for Component Based Design," University of California Berkeley, Electrical Engineering and Computer Sciences, Ph.D. Thesis, 2002.

[16] G. Nordstrom, J. Sztipanovits, G. Karsai, and A. Ledeczi, "Metamodeling - Rapid Design and Evolution of Domain-Specific Modeling Environment (ECBS'99)," *Proceedings of IEEE Conference and Workshop on Engineering of Computer-Based Systems*, Nashville, TN, April 1999, pp. 68–74.

[17] A. W. Brown and K. C. Wallnau, "Engineering of Component-Based Systems," *Proceedings of 2nd IEEE International Conference on Engineering of Complex Computer Systems*, 1996, pp. 414–422.

[18] A. C. Wills, "Modeling for Component Based Development," *Proceedings of European Conference on Object-Oriented Programming (ECOOP'98)*, 1998.

[19] B. Morel and P. Alexander, "SPARTACAS: Automating Component Reuse and Adaptation," *IEEE Transactions on Software Engineering*, Vol. 30, No. 9, September 2004, pp. 587–600.

[20] P. Alexander, *System-Level Design with Rosetta*, New York: Elsevier Science, 2006.

[21] Y. Chen and B. Cheng, "Facilitating an Automated Approach to Architecture-Based Software Reuse," *Proceedings of the 12th IEEE International Conference of Automated Software Engineering*, November 1997, pp. 238–245.

[22] M. Shaw, R. DeLine, D. Klein, T. Ross, D. Young, and G. Zelesnik, "Abstractions for Software Architecture and Tools to Support Them, "*IEEE Transactions on Software Engineering*, Vol. 21, No. 4, April 1995, pp. 314–335.

[23] J. Q. Ning, "Component-Based Software Engineering (CBSE)," *Proceedings of the 5th International Symposium on Assessment of Software Tools (SAST'97)*, Washington D.C., April 1997.

[24] Virtual Socket Interface Alliance, http://www.vsi.org.

[25] OCP International Partnership Association, Inc., http://www.OCPIP.org.

[26] Spirit Consortium, "SPIRIT Schema Working Group," http://www.spiritconsortium.org/.

[27] C. K. Lennard and E. Granata, "The Meta-Methods: Managing Design Risk During IP Selection and Integration," *Proceedings of European IP Conference*, November 1999.

[28] C++-Front-End, Edison Design Group, http://edg.com/cpp.html.

[29] M. Moy, F. Maraninchi, and L. Maillet-Contoz, "An Extraction Tool for SystemC Descriptions of Systems-on-a-Chip," *Proceedings of ACM International Conference on Embedded Software (EMSOFT'05)*, 2005, pp. 317–324.

[30] FZI - Microelectronic System Design, "KaSCPar: Karlsruhe SystemC Parser Suite," www.fzi.de/KaSCPar.html.

[31] W. Snyder, "SystemPerl," http://www.veripool.com/systemperl.html.

[32] D. Berner, H. D. Patel, D. A. Mathaikutty, and S. K. Shukla, "SystemCXML: An Extensible Systemc Frontend Using XML," *Proceedings of Forum on Design and Specification Languages (FDL'05)*, September 2005.

[33] D. Berner, H. D. Patel, D. A. Mathaikutty, and S. K. Shukla, "Automated Extraction of Structural Information from SystemC-Based IP for Validation," *Proceedings of the 6th International Workshop on Microprocessor Test and Verification (MTV'05)*, November 2005, pp. 99–104.

[34] M. Metzger, F. Bastien, F. Rousseau, J. Vachon, and E. M. Aboulhamid, "Introspection Mechanisms for Semi-Formal Verification in a System-Level Design Environment," *Proceedings of the 17th IEEE International Workshop on Rapid System Prototyping (RSP'06)*, June 2006, pp. 91–97.

[35] B. DuCharme, *XML: The Annotated Specification*, Upper Saddle River, NJ: Prentice Hall, 1998.

[36] Doxygen Team, "Doxygen," http://www.stack.nl/ dimitri/doxygen/.

[37] Apache Software Foundation, "Xerces C++ Validating XML Parser," http://xml.apache.org/xerces-c/.

[38] J. Lapalme and G. Nicolescu, "A New Efficient EDA Tool Design Methodology," *ACM Transactions on Embedded Computing System*, Vol. 5, No. 2, May 2006, pp. 408–430.

[39] Microsoft, Microsoft .NET, http://www.microsoft.com/net/.

[40] Microsoft, Microsoft Visual Studio C#, "http://msdn.microsoft.com/vstudio/express/visualcsharp/.

[41] P. Mishra and N. Dutt, *Processor Description Languages*, San Francisco, CA: Morgan Kaufmann Publishers, 2008.

[42] M. Freericks, "The nML Machine Description Formalism," Computer Science Department of TU Berlin, Tech. Rep. TR SM-IMP/DIST/08, 1993.

[43] G. Hadjiyiannis, S. Hanono, and S. Devadas, "ISDL: An Instruction Set Description Language for Retargetability," *Proceedings of Design Automation Conference (DAC'97)*, June 1997, pp. 299–302.

[44] A. Halambi, P. Grun, V. Ganesh, A. Khare, N. Dutt, and A. Nicolau, "EXPRESSION: A Language for Architecture Exploration Through Compiler/Simulator Retargetability," *Proceedings of the Design, Automation and Test in Europe (DATE'99)*, March 1999.

[45] V. Zivojnovic, S. Pees, and H. Meyr, "LISA — Machine Description Language and Generic Machine Model for HW/SW Co-Design," *Proceedings of the IEEE Workshop on VLSI Signal Processing*, 1996, pp. 127–136.

[46] A. Adir, E. Almog, L. Fournier, E. Marcus, M. Romon, M. Vinov, and A. Ziv, "Genesys-Pro: Innovations in Test Program Generation for Functional Processor Verification," *IEEE Design and Test of Computers*, Vol. 21, No. 2, 2004, pp. 84–93.

[47] P. Mishra and N. Dutt, "Automatic Functional Test Program Generation for Pipelined Processors Using Model Checking," *Proceedings of 7th IEEE International Workshop on High Level Design Validation and Test (HLDVT'02)*, 2002, pp. 99–103.

[48] S. Ur and Y. Yadin, "Micro Architecture Coverage Directed Generation of Test Programs," *Proceedings of the 36th ACM/IEEE Conference on Design Automation (DAC'99)*, 1999, pp. 175–180.

[49] M. Benjamin, D. Geist, A. Hartman, Y. Wolfsthal, G. Mas, and R. Smeets, "A Study in Coverage Driven Test Generation," *Proceedings of the 36th Design Automation Conference (DAC'99)*, June 1999, pp. 970–975.

[50] P. Mishra and N. Dutt, "Functional Coverage Driven Test Generation for Validation of Pipelined Processors," *Proceedings of the Conference on Design, Automation and Test in Europe (DATE'05)*, 2005, pp. 678–683.

[51] A. Dingankar, D. Mathaikutty, S. Kodakara, S. Shukla, and D. Lilja, "MMV: Metamodeling Based Microprocessor Validation Environment," *Proceedings of IEEE International High Level Design Validation and Test Workshop (HLDVT'06)*, November 2006, pp. 143–148.

[52] L. Bordeaux, Y. Hamadi, and L. Zhang, "Propositional Satisfiability and Constraint Programming: A Comparative Survey," *ACM Computing Surveys*, Vol. 38, No. 4, 2006.

[53] S. S. Muchnick, *Advanced Compiler Design and Implementation*, San Francisco, CA: Morgan Kaufmann Publishers, 1997.

[54] H. Patel and S. Shukla, "Model-Driven Validation of SystemC Designs," *Proceedings of the Design Automation Conference (DAC'07)*, 2007, pp. 29–34.

[55] A. Habibi and S. Tahar, "Design for Verification of SystemC Transaction Level Models," *IEEE Transactions on Very Large Scale Integration Systems*, Vol. 14, No. 1, 2006, pp. 57–68.

Part I

Design Reuse

Chapter 4

A Metamodel for Component Composition

The MCF design flow is shown in Figure 4.1, where the user begins by creating a component composition (CC) model using the component composition language. Upon conformance to the various constraints and completion of the model, it is converted to an intermediate representation (XML-IR), which is given as input to the introspective composer (IComp). On the other hand, we have an IP library of compiled C++ objects from which the metalevel information (structure, timing, and so forth) is extracted through automated reflective mechanism and expressed in XML. The extracted metainformation on the IPs in a library is also given as input to the introspective composer. It is responsible for constructing executable models by querying the reflected metainformation and selecting the appropriate IPs that are plugged into the component composition model. The main focus of this chapter is on the component composition language.

Component Composition Language

A visual language (CCL) for component composition is built using the metamodeling paradigm of the generic modeling environment (GME) [1]. It captures the syntax and interaction rules (semantics) of the visual composition language into a *metamodel*. The metamodel is developed as class diagrams using the Unified Modeling Language (UML) [2]. CCL allows a designer to visually construct abstract architectural templates conforming to the underlying semantics of the metamodel. A component composition model of a system design is also called an architectural template of the system.

Model Constraints

Modeling time constraints are enforced on objects and connections in a component composition model. The constraints are expressed as Boolean predicates using the Object Constraint Language (OCL) [3] and are captured as a part of the metamodel. These constraints check for the wellformedness of the modeling practice and enforce a correct-by-construction approach for modeling.

Analysis Engines

We develop three different interpreters as plug-ins for our CCF, which performs various kinds of analysis to detect system-level inconsistencies, type mismatches, and interface incompatibilities encountered during modeling. The first engine is a consistency checker that performs static analysis of the component composition model. It is similar to a "lint" checker for a hardware description language. The second engine is meant for datatype checking at the interface level and enforces that the data communicated between the different components are consistent. We developed algorithms to resolve the type-mismatch at the interface by type assignment such that there is no loss of data. Finally, a type propagation engine derives the types for a partially specified interface of a component based on architectural connections and the properties of the interacting components. Type inference algorithms have been developed for this purpose.

Translator

This engine translates the CC model into an intermediate XML [4] representation that will be given as input to the downstream tools. Note that this capability allows MCF to take an alternate route for an input specification without employing the visual language. The XML representation of the model can be imported into the modeling framework for visualization and model analysis, which does not limit the user to the visual language.

TCCL is a textual language that maps the visual constructs in the CCL to textual constructs. It is developed such that it has a one-to-one correspondence with the CCL. It is a way to address the scalability issue with the visual modeling framework and allows the designer to specify a textual architectural template of the system design. This template is converted into the XML-IR through a simple XML parser and imported into the modeling framework for visualization and application of the various constraints and checkers.

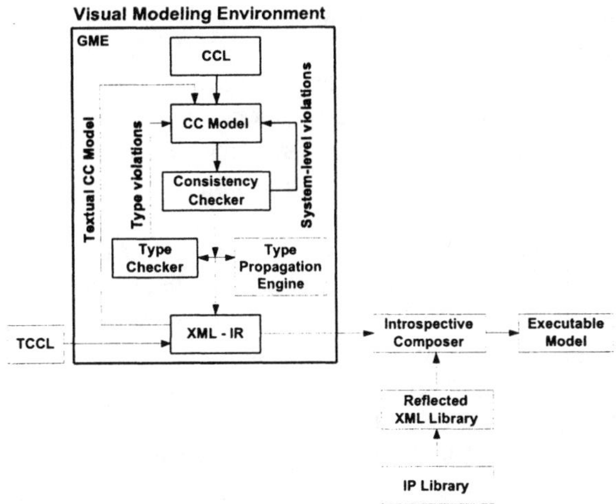

Figure 4.1 MCF design methodology.

4.1 CC LANGUAGE, METAMODEL, AND MODEL

MCF is used to describe an architectural template of a system design that will be instantiated with SystemC IPs from an IP library to create executable specifications. It is built on top of GME [1] and captures the syntax and semantics of a visual composition language (CCL) into an abstract notation called the *CC metamodel* (CCMM). It allows different design templates that facilitate a description in terms of component instantiations, characterization, and their composition that we call the *CC model* (CCM). The CCMM is developed using UML class diagrams and OCL constraints which are translated into visual syntax by GME. The visual design environment allows a modeler to construct models conforming to the underlying semantics of the CCMM.

4.1.1 Component Composition Language (CCL)

In order to enable SystemC-based IP integration, the CCL of MCF must reflect the hardware-related features allowed through SystemC. We show the mapping between the CCL of MCF and the SystemC language, in order to shed light on the underlying modeling constructs provided by CCL. MCF facilitates the creation of components and media at the architectural level, where components are computational blocks and the media are used for communication. The component instantiations that the language allows are leafs, hiers, and transactors. A *leaf* component is a representation of a hardware block without embedding other components. A *hier* component is a hierarchical hardware block that embeds other leaf or hier components. The structural interface of these component instantiations depends on the level of abstraction.

The component composition language has both RTL and TL abstractions similar to SystemC, which are captured through the metamodeling concept *aspect* in GME. A leaf component at the RTL abstraction in MCF, is described using an `SC_MODULE` instantiation in SystemC. It is allowed to have a set of input, output, and clock ports that map to the `sc_in⟨⟩`, `sc_out⟨⟩`, `sc_in_clk` and `sc_out_clk` constructs in SystemC. These SystemC constructs are used to describe the input, output, and clock ports. The RTL components in MCF communicate with each other through connections that map to the `sc_signal⟨⟩` construct in SystemC used to express signal-level communicating. SystemC does not provide a straightforward way of (de)-multiplexing ports. Demultiplexing is facilitated by reusing the same signal instance and multiplexing through sharing common signals across the communicating ports. This results in contention scenarios, which require the IP to have resolution logic to handle these scenarios. Furthermore, multiplexing and demultiplexing are common approaches to propagate the same data across IPs. Therefore to enable multiplexing and demultiplexing at the RTL abstraction, MCF provides mergers and splitters as communication media. A *merger* medium is a configurable MUX with n input ports, one output port, and m selection/driver ports. A *splitter* medium is a configurable DEMUX with one input port, n output ports, and m selection/driver ports.

At TL, MCF allows the modeler to describe interface skeletons, which are contained by one or more media. The interface skeleton maps to abstract classes in SystemC that derive from `SC_INTERFACE` construct. The media and other TL components have interface ports for communication that map to the `SC_PORT⟨⟩` construct in SystemC. The different media at TL map to channel implementations in SystemC that derive from the `SC_CHANNEL` construct.

It is a common practice to use specialized communication models to enable component interaction, such as buses and switches. In large systems, communication inevitably becomes a bottleneck and on-chip bus configurations and protocols significantly improve overall system performance. Therefore, we provide the capability to create bus media at the template-level in MCF. A *bus* medium is used to structurally portray contention and arbitration interfaces. In SystemC, a TL bus model is described using the SC_CHANNEL construct. Designing arrays of interconnected processors for multiprocessor SoC platforms has led to the evolution of a design methodology based on a new paradigm called NoC [5]. The common characteristics of a NoC-based system are that the processor and storage cores communicate with one another through intelligent switches. To cater for these kinds of architectures, we introduce a *switch* medium. A *switch* facilitates both point-to-point connections as well as complex message-passing protocols. At SystemC, a switch implementation at RTL is an SC_MODULE instantiation and at TL it is a SC_CHANNEL instantiation.

The concept of transactor has been proposed in SystemC to allow a TL-RTL mixed simulation [6]. A transactor works as a translator from a TL function call, to an RTL sequence of events (i.e., it provides the mapping between transaction-level requests, made by TL components, and detailed signal-level protocols on the interface of RTL IPs). Therefore, MCF also provides a mixed abstraction, where part of the structural interface of a component is at RTL and the other part is at TL. A component that exists at this abstraction in MCF is called a *transactor*. In MCF, the transactors specified at the architectural template are automatically synthesized during IP composition and are not subjected to IP selection.

A component in MCF is also allowed to contain other components and media, which enable hierarchical components. In SystemC, hierarchical descriptions are allowed by having multiple SC_MODULEs within a SC_MODULE. SystemC allows for different TL abstractions, MCF only captures the TL abstractions that are structurally distinguishable. Therefore, the TL capability provided through the CCL is called the programmer view (PV) of the abstraction [6].

SystemC allows for polymorphic modules and channels using C++ templates. A polymorphic module/channel can render its ports or/and parameters generic by binding them to type variables. It can also have polymorphic submodule instantiations, which indeed render their internals (port or/and parameters) polymorphic. MCF allows for a generic component/medium through partial specification, where it has ports or/and parameters that are UN. This implies that the types of the ports or/and parameters are not specified. MCF tries to infer these types during type propagation by looking at the architectural connections and the static properties. It can also be inferred during IP selection, when an IP implementation is used to instantiate the partially specified component/medium.

Note that MCF's RTL abstraction can be easily mapped to other hardware description languages such as VHDL [7], Verilog [8], and SystemVerilog [9].

4.1.2 Component Composition Metamodel (CCMM)

CCMM is represented as $CCMM = \langle E, R, E_C, R_C, A \rangle$ where E is a collection of entities and R is a collection of relationships that relate two objects. E_C is a collection of constraints described on the entities, whereas R_C is a collection of constraints on the relationships in the metamodel. Real-world objects are often represented and analyzed in different ways (e.g., multiple views). Similarly, a metamodel can also have multiple aspects, where each aspect reveals a different subset of entities and their relations. A is the set of aspects that define the possible visualizations for the CCMM.

The modeler instantiates an entity to characterize it by specifying its attributes and internals. The modeler instantiates a relationship to associate two or more instances of an entity or entities. The association is visually specified by connecting them.

Definition 1. *The allowable characterizations and possible containments of an entity and the relations that it can participate in, defines its static-requirement, which is fulfilled by the modeler and enforced by the metamodel.*

We discuss a few entities, relations, entity constraints, relation constraints, and aspects defined in the CCMM in this section. The complete description of the CCMM is provided in [10]. The two main entities in the CCMM are LEAF and CHANNEL, which belong to E. In Figure 4.2, we show the class diagram of a LEAF entity. Note that in the following figures, a triangle means *inheritance* and a black dot represents a *connection stream*, which has two links to denote the source (src) and destination (dst) of the connection and the third link denotes the <<Connection>> construct associated with the stream. The link with a black diamond at one of its ends means *containment*, where the object at the other end (child) has a containment relation with the object at the diamond-shaped end (parent).

The LEAF entity acts as a containment of four kinds of objects, namely IOPort, RWPort, CLKPort, and Parameter. The IOPort does not have an internal structure and is defined using the construct <<Atom>>. It acts as a placeholder for two specialized type of ports, namely INPUTPORT and OUTPUTPORT. These represent the input and output ports of an RTL component. The characteristics of such a port are captured through attributes, namely *portname*, *datatype*, and *size*, which are specified in the format (attribute name : attribute type). The other requirements of the IOPort object are expressed through its relationships. The IOConnection <<Connection>> is one such relation that associates two instances of the LEAF entity and belongs to R. This relation is used to portray the interaction between two RTL components via their input-output ports.

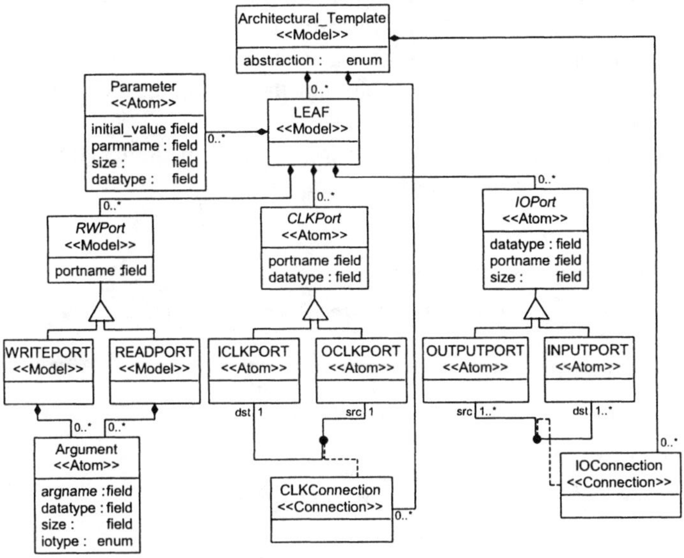

Figure 4.2 UML diagram of a LEAF object.

The second type of object that a LEAF can contain is RWPort. It is of two types READPORT and WRITEPORT that describes the read-write function calls of a TL component. These objects have an internal structure that allows the object to act as a container for objects of type Argument and therefore defined using the <<Model>> construct. These internals capture the static-requirements of read-write function calls, namely function arguments and return type. The attributes that characterize an Argument are *parmname*, *datatype*, and *iotype*.

The third object that a LEAF can contain is a CLKPort, which is also of two types, namely ICLKPORT and OCLKPORT, which portray the clocking ports of an RTL/TL component. The CLKConnection <<Connection>> that belongs to R is used to relate the clock ports of two instances in the template. Finally, the Parameter entity is defined using an <<Atom>> construct. It is used to represent the configurable factors of a component. The characteristics of a Parameter are captured through attributes, namely *parmname*, *datatype*, and *initial_value*.

At the top, we have an Architectural_Template entity that is defined as a <<Model>>. It is instantiated by the modeler for creating an architectural template of the system design. It has an *abstraction* attribute that captures the abstraction level at which the instantiated template will be created. The possible values for the enumerated attribute *abstraction* are *RTL*, *TL*, and *RTL_TL*. The Architectural_Template entity acts as the container for all possible components that a modeler is allowed to instantiate and the communication styles and legal interactions that are allowed.

We illustrate a few of the constraints defined in the CCMM that enforce metamodeling rules on a LEAF instantiation.

Listing 4.1 Constraint on design abstractions.

```
// Attached to the Architectural_Template <<Model>>
// Checked when a LEAF is internally instantiated with IOPORT or RWPORT

if(self.abstraction = "RTL") then
  // Enforces the constraint for a LEAF
  self.parts("LEAF")->forAll(obj:LEAF | obj.parts(RWPort)->size() = 0)
  ...
  else if(self.abstraction = "TL") then
    // Enforces the constraint for a LEAF
    self.parts("LEAF")->forAll(obj:LEAF | obj.parts(IOPort)->size() = 0)
    ...
    else if(self.abstraction = "RTL_TL") then
      //Enforces the constraint for a LEAF
      self.parts("LEAF")->forAll(obj:LEAF | (obj.parts(RWPort)->size() = 0
                                   and obj.parts(IOPort)->size() <> 0))
      or
      self.parts("LEAF")->forAll(obj:LEAF | (obj.parts(RWPort)->size() <> 0
                                   and obj.parts(IOPort)->size() = 0))
      ...
      else
          // The abstraction is not specified
          false
    endif
  endif
endif
```

The construct **self** points to the entity on which the constraint is attached. The function **parts**() collects the child entities of a parent entity into a set and **size**() returns the size of the set. The construct **forAll** is similar to a for-loop in C/C++ and it iterates over the entities in a set.

Listing 4.1 shows a part the OCL constraint that enforces modeling abstractions on the instances of the LEAF entity. The constraint in Listing 4.1 is attached to the Architectural_Template entity and is evaluated when an entity (LEAF, HIER, and so forth) is instantiated into the template and characterized. If this constraint does not evaluate to a *true*, then a violation is flagged. This part of the constraint enforces the following: (1) When the Architectural_Template's abstraction is set to *RTL*, the LEAF instances are not allowed to have RWPort instantiations, (2) when set to *TL*, the LEAF instances are not allowed to have IOPort instantiations, and (3) when set to *RTL_TL*, a LEAF instantiation is allowed to either have RWPort or IOPort instantiations, but not both.

Listing 4.2 The constraint checks for type mismatch at an IOPort.

```
// Attached to the INPUTPORT <<Atom>>
// Checked when an OUTPUTPORT and an INPUTPORT are connected by IOConnection

self.connectedFCOs("src")->forAll(obj:OUTPUTPORT |
                        ((obj.datatype = self.datatype and obj.size = self.size)
                        or obj.datatype = "UN" or self.datatype = "UN")
```

The function **connectedFCOs**() collects the source and destination entities connected to an entity, unless explicitly asked to collect either. The **let-in** construct is used to create local definitions, similar to the let environment in a functional language. In Listing 4.2, we show the constraint that checks for type mismatch when connecting an OUTPUTPORT to an INPUTPORT. It basically enforces that the datatype attribute of the two connected IOPort objects must be identical. If the two differ, then either of them must be unspecified, which means the datatype attribute is UN.

Listing 4.3 The constraint checks for wellformedness for an RWPort.

```
// Attached to the READPORT and WRITEPORT <<Model>>s
// Checked when an instantiated ARGUMENT in READPORT/WRITEPORT is characterized through argname.

let I = self.parts(Argument) in
    I->forAll(obj1:Argument | let J = I.excluding(obj1) in
                        J->forAll(obj2:Argument | obj1.argname <> obj2.argname))
```

Some of the OCL constraints that are enforced on a LEAF instance at the TL abstraction are shown next. The constraint shown in Listing 4.3 checks for the wellformedness of a read/write call construction using the RWPort object. A function definition in C++ requires that all its function arguments be unique. This requirement holds during the construction of a read/write call using the RWPort object in our CCL. It is enforced using the constraint shown in Listing 4.3, which requires that each Argument object instantiated within a RWPort to have distinct *argname* attributes. This uniqueness feature is used later on to check whether two interacting TL components have a type-consistent interface (shown later in Listing 4.7).

Two of the other entities that belong to E on which most of the OCL constraints attached to a LEAF entity apply are the HIER and TRANSACTOR. The UML diagram of the HIER is very similar to the one in Figure 4.2 with a few additions. The HIER entity is an extension of the LEAF entity to support hierarchy, which is enabled through the concept of containment. A way to think about the HIER entity is the LEAF with an extra containment relation, which is a self-loop, implying that it contains itself. In reality, the HIER entity can contain a completely new architectural template within, which requires it to facilitate template creation. Therefore, the HIER entity is more complex than a simple extension to the LEAF entity, but is introduced into the CCMM to allow hierarchical components.

The TRANSACTOR entity has the same UML description and OCL constraints as the LEAF entity with the exception of the design abstraction constraint. This exception is shown in Listing 4.4 and it enforces that a TRANSACTOR instantiation can only exist in the RTL_TL abstraction.

Listing 4.4 The design abstraction constraint on a TRANSACTOR entity.

```
// Attached to the Architectural_Template <<Model>>
// Checked when a TRANSACTOR is instantiated

if(self.abstraction = "RTL" or self.abstraction = "TL")
then
    // Enforces the constraint for a TRANSACTOR
    self.parts("TRANSACTOR")->size() = 0
else
  if(self.abstraction = "RTL_TL") then
    true
  else
    false
  endif
endif
```

A transaction interface in SystemC is an abstract class that inherits from the `SC_INTERFACE` class. It contains the declaration of virtual functions that describes the skeleton of the possible transactions allowed through the interface. The interface can be implemented by one or more classes that inherit from `SC_CHANNEL` class. During this implementation, the interface functions are defined in terms of what data is exchanged as well as how the data is exchanged through the interface using them.

We employ the INTERFACE object to depict a transaction interface and the CHANNEL entity to portray its implementation using the UML description shown in Figure 4.3. The INTERFACE object is allowed to contain RWPort objects to describe the read and write transaction calls. The INTERFACE instantiations are collected by the Interface_List entity. The CHANNEL entity acts as a container for two types of ports, naming RWPort and CLKPort, therefore a CHANNEL instantiation is only allowed at the TL abstraction of the template. We capture "a channel implements one or more interfaces," using a containment and association relation. A CHANNEL instance is allowed to contain one or more INTERFACE instantiations and the RWPort objects in these INTERFACE instantiations are connected to the RWPort objects contained in the CHANNEL using the INTFConnection <<Connection>>. Therefore, the drag-and-drop of an INTERFACE instance ($intf$) into a CHANNEL instance (ch) and connecting their RWPort objects imply that ch implements $intf$.

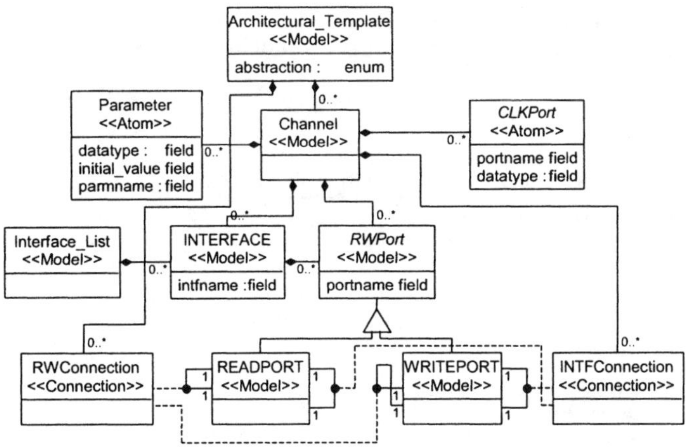

Figure 4.3 UML diagram of a CHANNEL object.

With such a nonsimple way to describe how a channel implements an interface, the modeler is certain to make mistakes. Therefore, we have a set of OCL constraints written to enforce a correct interface and channel description. Some of these constraints are shown later.

In Listing 4.5, we show the completeness constraint that checks each INTERFACE instance in a CHANNEL to see if the modeler has forgotten any of the necessary INTFConnection connections. If an RWPort (read/write function declaration) of an INTERFACE instance is not connected to an RWPort of the CHANNEL instance, then it would be interpreted as the implementation (function definition) is missing.

Listing 4.5 Completeness constraint on a CHANNEL entity.

```
// Attached to the CHANNEL <<MODEL>>
// Checked when a CHANNEL's INTERFACE is being instantiated and connected using INTFConnection

self.parts()->size() > 0 and
self.parts(READPORT)->forAll(obj:READPORT |
                              obj.connectedFCOs("dst",INTFConnection)->size() = 1)
and
self.parts(WRITEPORT)->forAll(obj:WRITEPORT |
                              obj->connectedFCOs("dst",INTFConnection)->size() = 1)
and
self.parts(INTERFACE)->forAll(obj:INTERFACE |
 (obj.parts(READPORT)->forAll(obj1:READPORT |
                              obj1.connectedFCOs("src",INTFConnection)->size() = 1))
 and
 (obj.parts(WRITEPORT)->forAll(obj1:WRITEPORT |
                              obj1.connectedFCOs("src",INTFConnection)->size() = 1))    )
```

Listing 4.6 Constraint to avoid duplicate INTERFACE instances.

```
// Attached to the CHANNEL <<MODEL>>
// Checked when an instantiated CHANNEL's INTERFACE is being instantiated and characterized

let
I = self.parts("INTERFACE") in
I->forAll(obj1:INTERFACE |
   let J = I.excluding(obj1) in
   J->forAll(obj2:INTERFACE |
     obj1.intfname <> obj2.intfname and
     (obj2.parts(READPORT)->forAll(obj3:READPORT |
          obj1.parts(READPORT)->forAll(obj4:READPORT | obj4.portname <> obj3.portname))) and
     (obj2.parts(WRITEPORT)->forAll(obj3:WRITEPORT |
          obj1.parts(WRITEPORT)->forAll(obj4:WRITEPORT | obj4.portname <> obj3.portname)))
   )
)
```

In Listing 4.6, we show the constraint that enforces that the same INTERFACE instance cannot be dropped more than once into the CHANNEL instance. The interfaces are uniquely identified by the *intfname* attribute associated with them. Two INTERFACE instances can have distinct *intfname* attributes, but internally the same RWPort instantiations. This means that two INTERFACE instances with identical transaction calls exist in the same CHANNEL, which is not allowed and this constraint catches such a violation.

In SystemC, TL components communicate through transaction ports that are hooked on to one or more channels. A transaction port allows the component to use read and write functions defined by the channel. These functions are declared by an interface. The interface is implemented by a channel, which provides the definition for the functions declared in the interface. Therefore, TL components connect to channels that implement the interface they use to perform their transactions. We allow a similar description capability through the CCMM to facilitate transaction-level depiction. A TL LEAF instance interacts with other TL LEAF instances using CHANNEL instances containing the necessary INTERFACE instantiations. This implies that for two TL LEAF instances L_1 and L_2 interacting through a CHANNEL instance ch, the RWPort ports of L_1 and L_2 are also contained in ch. These RWPort ports of ch are connected to RWPort ports of the INTERFACE instances contained in ch through an INTFConnection <<Connection>>. Therefore to represent the interaction between L_1 and ch as well as L_2 and ch, we provide RWConnection <<Connection>>. This connection is understood as a transaction port and allows L_1 and L_2 to use the transaction functions defined by ch.

Note the following: (1) INTFConnection is used to connect a READPORT or WRITEPORT object in a CHANNEL instance to a READPORT or WRITEPORT object in an INTERFACE instance, where this connection depicts that the CHANNEL instance implements the INTERFACE instance by defining its functions. (2) RWConnection is used to connect a READPORT or WRITEPORT object in a LEAF instance to a READPORT or WRITEPORT object in a CHANNEL instance, where this connection depicts that the LEAF instance uses the read or write function defined by the CHANNEL. In order to enforce the correct usage of the RWConnection and INTFConnection <<Connection>>s, we have constraints attached to them.

For a CHANNEL instance, we check for interface compatibility at the datatype-level along the RWConnection and INTFConnection <<Connection>>s. The constraint that checks for compatibility on READPORT objects is shown in Listing 4.7 and the same constraint has a part for WRITEPORT objects. In order to formulate this constraint, we employ the uniqueness feature of the Argument object.

Some of the other entities not discussed so far are MERGER, SPLITTER, BUS, and SWITCH. The MERGER and SPLITTER instantiations are only allowed at the RTL abstraction of the template. Their instantiations represent MUX and DEMUX components at the hardware level. They are needed in the CCMM to allow multiplexing and demultiplexing of connections, since the IOPort objects of a LEAF instance are only allowed point-to-point connections. One of the static-requirements of a MERGER (MUX component) is that all its input multiplexing ports and the output multiplexed port must have type consistent datatype requirements. This requirement is enforced through OCL constraints and similar constraints are enforced on a SPLITTER instantiation based on its static-requirement. The BUS entity is provided to allow bus-style interaction, where a bus port is allowed to be driven by multiple ports. The SWITCH entity is provided to allow different topological interactions such as mesh-based, ring-based, tree-based, and so forth. The static properties of both these entities are also enforced through constraints.

Listing 4.7 Constraint for interface compatibility of a CHANNEL instance.

```
// Attached to the CHANNEL <<Model>>
// Checked when two READPORTs are connected using the connection RWConnection/INTFConnection

let I = self.parts(READPORT) in
  let J = I.connectedFCOs() in
    J->forAll(obj1:READPORT |
    let K = obj1.parts(Argument) in
      let L = I.parts(Argument) in
      K->size() = L->size() and
      K->forAll(obj2:Argument |
          L->exists(obj3:Argument | (obj3.argname  = obj2.argname) and
                                  ((obj3.datatype = obj2.datatype and obj3.size = obj2.size) or
                                    obj3.datatype = "UN" or obj2.datatype = "UN" ))
      )
  )
```

When modeling a system, it is important that the requirements be captured in a unified set of models. This unified model set is then partitioned based on the system's overall behavior, which facilitates an examination of the system focusing on specific aspects. The model design space is partitioned into aspects, where each aspect encloses a part of the design space within a particular model. Depending on the aspect chosen, certain modeling parts and/or relationships may or may not be allowed. More than one aspect can exist within a particular model. Aspect partitioning is also a way to handle model complexity with visual modeling tools. Some of the aspects developed for CCMM are shown here:

$$A = \{\text{DesignView}, \text{Point2PointView}, \text{ChannelView}, \text{MixedView}\}$$

Point2PointView is meant for RTL-specific visualization, *ChannelView* is for TL-specific visualization and *MixedView* is for RTL_TL visualization. The aspect *DesignView* allows in visualizing the complete architectural template of the system as a flat model.

4.1.3 Component Composition Model (CCM)

A CCM is the instantiation, characterization, and connection of entities from the CCMM to create an architectural template of the system. It is represented as $CCM(a) = \langle I, I_C, C, C_C \rangle$. I is the set of instantiations of the different entities in the CCMM. I_C collects the activated constraints on the instances. We had illustrated some of the OCL constraints formulated as a part of the CCMM in the previous subsection. The constraint attached to an entity is activated as soon as the entity is instantiated and configured or characterized. A constraint violation will undo the incorrect modeling step performed by the modeler, thereby promoting a modeling process that is correct-by-construction. C is a set of connections, where a connection associates two objects through a relationship defined in the CCMM. We use connections in our CCM primarily for modeling different kinds of interactions between instances. C_C collects the activated constraints on the connections, which are also enforced by CCMM and checked at modeling time. These constraints are activated when the instantiated entities are connected using the <<Connection>>s on which they are attached. Finally, the variable "a" is initialized with a value from A giving the CCM a particular visualization.

Every instance of an entity is associated with a *name* that is predefined with the name of the entity being instantiated. The user characterizes this instance with a unique name as a part of its description (C1, C2, and so forth in Figure 4.4(b)); unless an already existing instance is being reused. The instances in the CCM are grouped into three distinct categories, namely components (C_E), medium (M_E), and clock generators. C_E is the collection of LEAF, HIER, and TRANSACTOR instances, whereas M_E is a collection of MERGER, SPLITTER, CHANNEL, BUS, and SWITCH instances. We provide examples to explain the CCM creation.

During modeling, say the user begins by instantiating the LEAF entity to describe a component and proceeds to describe the internals. Based on whether the user wants an RTL or TL description, the internals would be described by instantiating one or more IOPort or RWPort objects. These internals are then characterized by filling in their attributes, which describe the type of data communicated through these ports. Furthermore, the component can be clocked by instantiating CLKPort objects. Even though the LEAF instance in the CCM can contain three different types of objects, at modeling time it is not allowed to have both IOPort and RWPort objects, since a component can either be at RTL or at TL (Listing 4.1 shows this constraint). For an RTL SystemC description shown in Figure 4.4(a), we show the RTL CCM created using the CCL in Figure 4.4(b). The blocks with the centered "L" are LEAF instances with IOPort objects and CLKPort objects. The dashed lines show CLKConnection instantiations and the other lines show IOConnection instantiations.

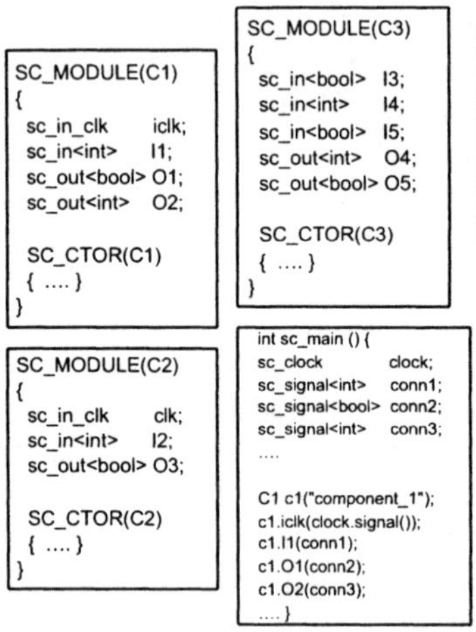

(a) RTL SystemC model.

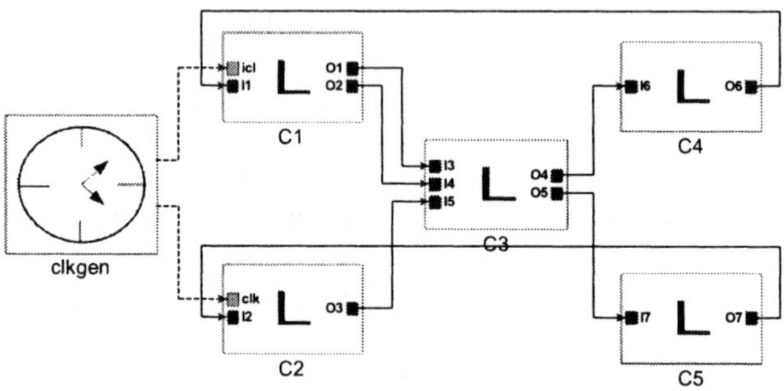

(b) RTL CCM.

Figure 4.4 (a, b) An abstract RTL example.

Each IOPort instantiation is characterized by their attributes, namely *port-name* (id), *datatype* (dt), and *size* (sz). Let T be the type-system, then $dt \in T$ and $sz \in N$. The *size* attribute is used to capture the bitwidth of the port. Consider the

INPUTPORT instance I_3 of LEAF instance C_3 shown in Figure 4.4(b). Characterization of I_3 using the SystemC description in Figure 4.4(a) and the allowable types in MCF in Figure 4.7 is shown here:

$$f_{id}(\text{INPUTPORT}, I_3),\ f_{dt}(I_3,\text{bool}) \text{ and } f_{sz}(I_3,8)$$

The functions f_{id}, f_{dt} and f_{sz} are used to assign the attributes with user-defined values. The function f_{dt} takes two arguments, namely an object and a value and places the value in the *datatype* attribute of the object. At modeling time, the user selects the instantiated entity that needs to be characterized within the visual framework and textually specifies the values for its attributes.

During characterization, the modeler specifies a value for the *datatype* attribute of an IOPort object. This value must be a legal C++/SystemC datatype and a part of the allowable types in MCF. The complete set of C++ and SystemC datatypes supported by MCF are provided in Figure 4.7. Note that for all C++ datatypes the *size* attribute is predefined and follows from Figure 4.2 on the left. Some of the SystemC types are configurable with respect to its bitwidth as shown in Figure 4.2 on the right and that capability is enabled through the *size* attribute. Attaching a datatype to an IOPort object is referred to as providing a *type-binding*. The datatype specification of every IOPort object of a LEAF instance is called its *type-contract*. The modeler also has the choice of not specifying the *datatype* attribute of an IOPort object. In this case, the *datatype* attribute is assigned with UN, which implies that the datatype is not known at this point.

Definition 2. *The type-contract of an instance at the RTL abstraction describes the type-bindings of all its IOPort objects.*

At the TL abstraction, a LEAF instance is allowed to contain RWPort objects to depict read-write functions. A read/write function has a set of arguments, which are read from or written to during the transaction. A read function reads from its arguments, whereas a write function modifies its arguments. We enable this capability in our CCMM, by allowing each RWPort instantiation to contain Argument objects. An Argument object is characterized by their attributes, namely *argname*, *datatype*, and *iotype*. During characterization, the modeler specifies a value for the *datatype* attribute of an Argument object in an RWPort object. This value must be a legal C++/SystemC datatype and a part of the allowable types in MCF. Attaching a datatype to an Argument object is referred to as providing a *type-binding*. The datatype specification of every Argument instance in every RWPort object of a LEAF instance is called its *type-contract*. In case the modeler does not want to specify a datatype, the *datatype* attribute is assigned with UN.

Definition 3. *The type-contract of an instance at the TL abstraction describes the type-bindings of all the Argument objects in all its RWPort objects.*

Definition 4. *An instance is partially typed, when the type-contract consists of at least one UN type-binding. An instance is untyped when all its type-bindings in the type-contract are UN.*

Consider the example of a producer and consumer communicating through a FIFO. Its TL description in SystemC is shown in Figure 4.5(a). The corresponding CCM is shown in Figure 4.5(b) with two LEAF instances, namely *Producer* and *Consumer* that interact through a CHANNEL instance *FIFO*. Note that both the *Producer* and *Consumer* connect to the *FIFO* using RWConnection instantiations.

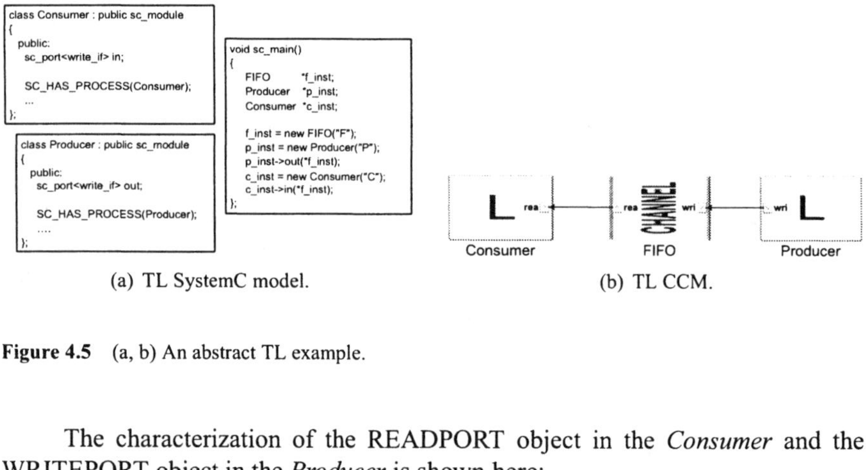

(a) TL SystemC model. (b) TL CCM.

Figure 4.5 (a, b) An abstract TL example.

The characterization of the READPORT object in the *Consumer* and the WRITEPORT object in the *Producer* is shown here:

f_{id}(READPORT, read)
  f_{id}(read.Argument, Data)
    f_{dt}(Data, *char*), f_{sz}(Data,8), and f_{iotype}(Data, output)
f_{id}(WRITEPORT, write)
  f_{id}(write.Argument, Data)
    f_{dt}(Data, *char*), f_{sz}(Data,8), and f_{iotype}(Data, input)

In this example, we instantiate a READPORT $read$ with an Argument object $Data$ of type character. The *Data* is read from during the transaction, for which the *iotype* attribute is set to output. We also instantiate a WRITEPORT $write$ with an Argument object $Data$ of type character. It is modified during the transaction, for which the *iotype* attribute is set to input. A function call could be bidirectional, in which case some of its arguments are read from and some others are written into, in which case we denote the direction of these arguments through the *iotype* attribute. This characterization follows from the interface description in Figure 4.6(a). The interface that the FIFO implements in the producer-consumer model is shown in Figure 4.6(a).

```
class write_if : virtual public sc_interface
{
  public:
  virtual void write(char Data) = 0;
};
```

```
class read_if : virtual public sc_interface
{
  public:
  virtual void write(char Data) = 0;
};
```

```
class FIFO : public sc_channel, public write_if, public read_if
{
  public:
  FIFO(sc_module_name name) : sc_channel(name) {}

  void write(char Data) { .... }
  void read(char Data) { .... }
}
```

(a) SystemC description.

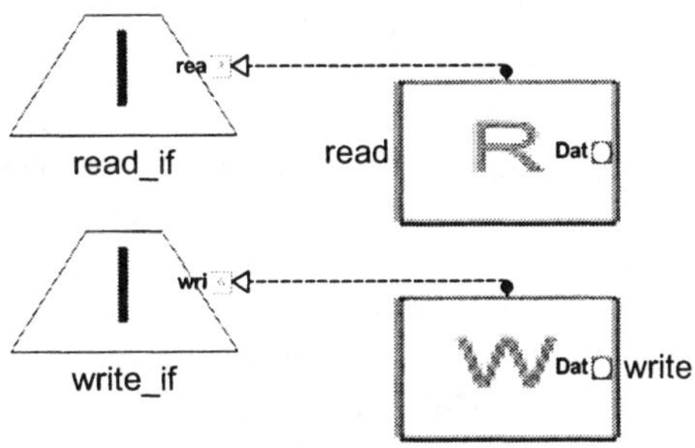

(b) CCM of FIFO internals.

Figure 4.6 (a, b) TL interfaces.

Note that in order to implement the *read_if* and *write_if* interfaces, the *FIFO* first inherits from them and then provides the definition for the *read* and *write* virtual functions. We mimic this description by using INTERFACE instances to describe both the *read_if* and *write_if* interfaces. These INTERFACE instances are then dragged and dropped into the *FIFO* CHANNEL instance, which maps to the FIFO inheriting the interfaces in the SystemC description. Then, we connect the RWPort objects in these INTERFACE instances (read_if.read and write_if.write) with the RWPort objects in the FIFO CHANNEL instance (FIFO.read and FIFO.write) using INTFConnection instantiations. This maps to the FIFO defining the read and write function in the SystemC description. The internals of the FIFO CHANNEL instance in the CCM is shown in Figure 4.6(b). The parallelograms with a centered I are INTERFACE instances that represent the read_if and write_if interfaces. The dashed lines represent INTFConnection instantiations. The R and W boxes represent the RWPort objects of the CHANNEL instance.

4.2 CC ANALYSIS AND TRANSLATION

Developing a CCL using a UML-based visual language brings forth the advantages associated with visualization and object-oriented features. However, the flipside is that the designer's intent and the language expressiveness must be carefully and explicitly captured for such a framework. Precisely or formally capturing these requirements, disallow any unwanted or illegal description through the CCL. Some of the requirements of the CCL are formalized and enforced using OCL. Some other requirements entail analysis of the complete model through inference techniques and we enforce these through checkers. We have implemented three different engines, first, a consistency checker that disallows any discrepancies in the CCM and second, a type-checker that looks for type mismatches between interacting components in the CCM. We also developed a type propagation engine that performs type derivations through type inference and type checking. Finally, we have a translator that converts the visual CCM into an intermediate representation based on an XML schema. This XML representation of the model is processed by the downstream tools to enable IP composition.

4.2.1 Consistency Checking

Some of the HDL related consistency checks performed include whether connected components match in terms of port sizes, interface calls, whether unbounded ports or duplicate ports exist, and so forth. In this respect, the checker is similar to a linting tool for an HDL language. The components in the CCM are devoid of an implementation and therefore behavioral syntax and semantic checks cannot be performed by this checker. On the other hand, automatic selection of an implementation from the IP library requires the component to be annotated with sufficient metadata. For example, if the user wants to perform selection of IPs based on the nomenclature, then the components must have a unique identity that is enforced by the checker. If two components have identical names, then they are inferred as instances of the same component that is reused and checked for structural equivalence.

Definition 5. *Two instances are said to be structurally equivalent when their type-contracts are identical.*

Name	**Description**	**Size**
char	Character or small integer	1 byte
wchar_t	Wide character	2 byte
short int/short	Short integer	2 byte
int	Integer	4 byte
long	Long integer	4 byte
long long	Long long integer	8 byte
float	Floating point (FP) number	4 byte
double	Double precision FP number	8 byte
long double	Long double precision FP number	16 byte
string	String	N byte
bool	Boolean	1 byte
sc_bit	Single bit value : '0' / '1'	1 bit
sc_bv⟨N⟩	Vector of sc_bit values	N bits
sc_logic	Single bit value : '0' / '1' / 'X' / 'Z'	2 bits
sc_lv⟨N⟩	Vector of sc_logic values	N bits
sc_int⟨N⟩	Integer	N bits (max 64)
sc_bigint⟨N⟩	Big integer	N bits (max 512)
sc_uint⟨N⟩	Integer	N bits (max 64)
sc_biguint⟨N⟩	Big integer	N bits (max 512)

Figure 4.7 C++ and SystemC types supported by MCF.

A consistency checker is an important part of the CCF; it finds inconsistencies early in the specification that reduces the modeling time. Furthermore, some of the tedious tasks are also automated through this checker, such as naming the relation instances. We do not require the user to name the <<Connection>> instantiated for connecting the instances. It is tedious and an error-prone task, which at the RTL in SystemC maps to using `sc_signal⟨⟩` constructs for connecting the modules (Figure 4.4(a)). We automate the task of uniquely naming all the relation instances through the checker.

4.2.2 Type Inference

Systems described with a strong-typed language such as C++ requires components to be implemented using interfaces that are type compatible. While composing IPs in C++, manually ensuring such type compatibility results in significant design-time overhead. In this section, we discuss inference techniques that employ type-checking and type propagation to simplify the strong typing requirements of the target design language (C++).

Type-System

In order to allow for type-checking and type propagation, we capture the relevant types of the target language into a *type-system* that is given as input to the different checkers and engines. Figure 4.8 shows the MCF type-system.

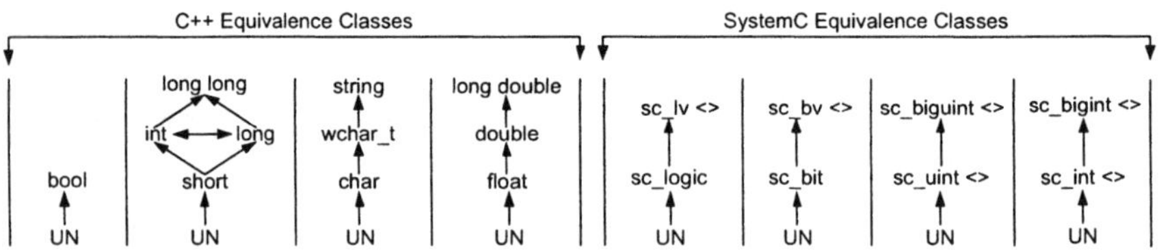

Figure 4.8 MCF type-system.

Definition 6. ***Type-system:*** *It is a directed acyclic graph, which represents the hierarchy for the basic types supported by the design domain.*

The type-system is a way of dividing the datatypes provided in the design domain into equivalence classes. The notion of equivalence is derived from the context of usage and an equivalence class contains types that are approximations of each other. For example, `char` and `string` belong to one class, whereas `int` and `long` belong to another class. This is because the type `string` is an expansion of the type `char`. The set of types from C++ and SystemC supported by MCF is shown in Figure 4.7. First, we distinguish between the C++ types and the SystemC types and then define equivalence classes in both. In Figure 4.8, the equivalence classes separated using vertical bars are shown.

Note that none of the C++ types are configurable in terms of their bit or byte size. However, some of the SystemC types are configurable such as `sc_int`⟨**N**⟩, where **N** is the number of bits. In order to capture the type bitwidth configurability, we provide an attribute *size* with all objects that has a *datatype* attribute in the CCMM. However, if the *datatype* attribute is characterized with a C++ type, then the *size* attribute has a predefined value that is enforced through a constraint. Part of the constraint with respect to the types `long` and `long int` attached to an IOPort object is shown in Listing 4.8. Currently, MCF's type-system consists of basic C++ and SystemC types and does not include composite or user-defined types.

Listing 4.8 Bitwidth constraint on C++ datatypes.

```
// Attached to the IOPORT <<Atom>>
// Checked when the datatype attribute of an IOPort is being characterized

self.parts(IOPort)->forAll(obj:IOPort |
 if (obj.datatype = "long" or obj.datatype = "long int")
  then obj.size = "32"
 else
   ...
 endif
)
```

Type-Checking

The type-system is used to perform two types of checking that are defined here:

Definition 7. ***Type equivalence:*** *Two types are said to be type equivalent if they map to the same node of the MCF type-system.*

Definition 8. ***Type subsumption:*** *A type t_i is subsumed by t_j if both belong to the same connected subgraph of the MCF type-system and there exists at least one path from t_i to t_j.*

The OCL constraints perform type equivalence checks, where the datatypes being matched are type equivalent if they map to the same node of an equivalence class. On the other hand, type subsumption provides more flexibility within the same equivalence class, for which the datatypes being matched, must satisfy Definition 8. If type t_i is subsumed by t_j, then this physically means an output port of type t_i can be legally connected to an input port of type t_j. A special case of type equivalence that arises from the configurability aspect of SystemC datatypes is handled as a type subsumption case during type-checking. This happens when the matched datatypes map to the same node during type-checking, but have different bitwidths. For example, let the two datatypes matched be `sc_int`$\langle 8 \rangle$ and `sc_int`$\langle 16 \rangle$. The type-checker resorts to type subsumption than type equivalence, even though it more closely follows the definition of the latter. Therefore, we make the following addendum to the type subsumption definition:

Definition 9. *A type $t_i\langle n \rangle$ is subsumed by $t_j\langle m \rangle$ if they map to the same node of the MCF type-system and $n < m$, where $n, m \in \mathbb{N}$.*

If the datatypes being matched belong to different equivalence classes then type-checking fails on both accounts. Currently, we do not support type-checking using the notion of subtyping. For example, one could consider the type `char` to be a subtype of the type `int`. This requires connecting the subgraphs of the different equivalence classes to arrive at a lattice with the top element being the most general type and the bottom element being UN. This construction may or may not be possible and is part of our future work. A type-lattice construction from the types used in the embedded software domain has been shown in the context of Ptolemy's work [11].

A component's interface description using our CCMM is dependent on the abstraction level and so are our checks. At RTL, type-compatibility boils down to checking whether output and input ports of the communicating components match in terms of datatype and bitwidth. At TL, type-checking resorts to verifying that the read/write calls match in terms of function signatures (return value, argument types, and number of arguments).

We perform type-checking during modeling using OCL constraints and during CC analysis using our type-checker. The type checks expressed as OCL constraints are attached to entities and relations. These at the RTL abstraction enforce that the ports of connected instances are *type equivalent* (Listing 4.2). Figure 4.9 illustrates an RTL component C_1 connected to C_2 in Ex1 and a bus media with an input port driven by output ports of components $master_1$ and $master_2$ in Ex2.

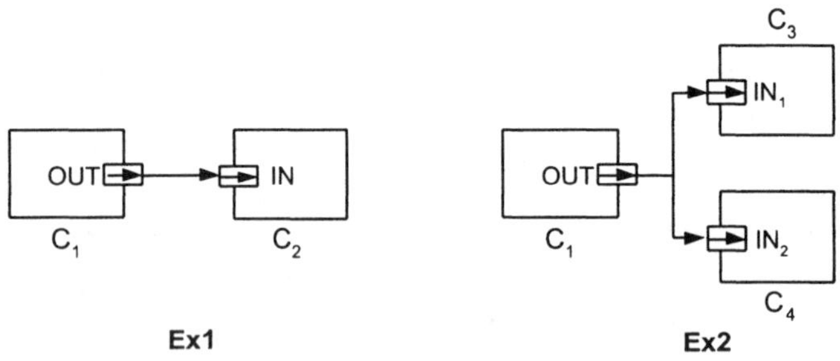

Figure 4.9 Examples that illustrate constraints.

The OCL constraints enforced by the framework checks are the following:

1. For Ex1: $(C_2.in.dt = C_1.out.dt)$ and $(C_2.in.sz = C_1.out.sz)$
2. For Ex2: $((M_1.p_1.dt = B.p_b.dt)$ and $(M_1.p_1.sz = B.p_b.sz))$ and $((M_2.p_2.dt = B.p_b.dt)$ and $(M_2.p_2.sz = B.p_b.sz))$

Constraint 1 enforces a rule on out and in that both these ports must have identical datatypes and bitwidths. This is an example of an *association-constraint* because it is activated as soon as the port out is connected to the port in. Constraint 2 enforces a rule on (p_1 and p_b) as well as (p_2 and p_b), which says that both components communicating through a common bus port must have their ports (p_1 and p_2) bound to the same type or unspecified. In Ex2, the ports p_1 and p_2 are not connected, but the Constraint 2 applies an association-constraint between them. Constraint 2 is an example of *static-requirement-constraint*. Some of the entities in the metamodel have a specific set of type properties that are mapped to static-requirement-constraints. In a very abstract sense, they can be thought of as constraints on association constraints. They constrain two different port-connections of an instance (that may or may not share ports) based on the properties of the instance. The OCL constraints catch association- and static-requirement-constraint violations at modeling time rather than have the type-checker catch these during analysis.

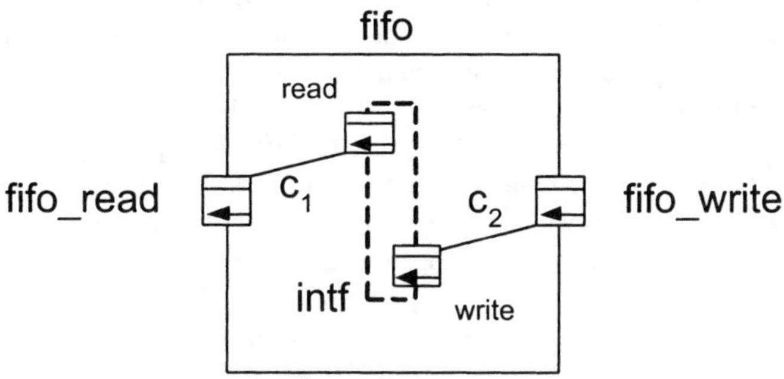

Figure 4.10 Fifo CHANNEL example.

The type-checking at the TL resorts to verifying that function signatures match without failing using OCL constraints. Consider the example of the *Fifo* CHANNEL instance shown in Figure 4.10 that interacts with a LEAF instance C_1. The connections c_1 and c_2 enforce that the Argument objects of the RWPort objects $fifo\_read$ and $num\_available$ are *type equivalent* to those of the RWPort objects $read$ and $size$. This is expressed as an OCL constraint in Listing 4.7. The Argument object are matched and type checked using their distinct *argname* attribute. Furthermore, the RWPort objects must match in the number of Argument objects. Let S denote the function signature, then the following must be true for the RWConnection instantiation c_1:

$|fifo\_read| = |read|$ and $S(fifo\_read) = S(read)$,
where,
$((fifo\_read.Data.dt = read.Data.dt)$ and $(fifo\_read.Data.sz = read.Data.sz))$ and
$((fifo\_read.Address.dt = read.Address.dt)$ and $(fifo\_read.Address.sz = read.Address.sz))$

where $|fifo\_read|$ is the number of Argument objects in the RWPort object $fifo\_read$. Note that a similar constraint is enforced on the RWConnection instantiation c_2.

The OCL constraints attached on the entities and relations in the metamodel are limited to type-checking based on type equivalence. Consider a SystemC example, where an adder's output of type `sc_int<16>` is connected to the input of a multiplier of type `sc_bigint<16>`. The activated OCL constraints will flag a violation; however the types can be equated through subsumption. Therefore, if the modeler wants to use the subsumption relation instead of type equivalence, then the failing of the OCL constraints serves as a notification that the user has connected two ports that are not type equivalent. The checking based on *type subsumption* is performed by our type-checker that derives the type relations from the MCF type-system. If a type relation cannot be derived for the two specified types, then a violation is reported by the checker.

The type-checking engine also tries to infer and propagate a type onto untyped/partially typed components in the template. The inference step utilizes *association-constraints* as well as *static-requirement-constraints* on the instances in the template to derive new type-bindings for an untyped/partially typed component. The propagation step assigns these type-bindings on the partially typed/untyped port(s) of the component. These steps are followed by a new round of type-checking to ensure that the assignment has not resulted in any inconsistencies. Before we provide the details of our inference algorithm, we explain the need for type inference at the architectural level in MCF.

Partial Specification

During the architectural specification in MCF, a component's interface is bound to a set of datatypes that is collected to create its contract. The contract describes type-bindings that the component must conform to during its interaction. MCF also allows a *partial architectural specification* to enable the description of flexible components, during which a component is made typeless by binding it to UN. On the other hand, if a component is bound to a type-contract, then its behavior is derived from its contract; thereby restricting its flexibility. Therefore, partially specified architectural templates provide an easy and fast specification process. If the modeler wants to specify a generic RTL adder, then all its ports have their *datatype* attribute characterized as UN. Consider the generic adder (Adder_2_1) shown in Figure 4.11, where based on the usage, the port bindings are derived differently.

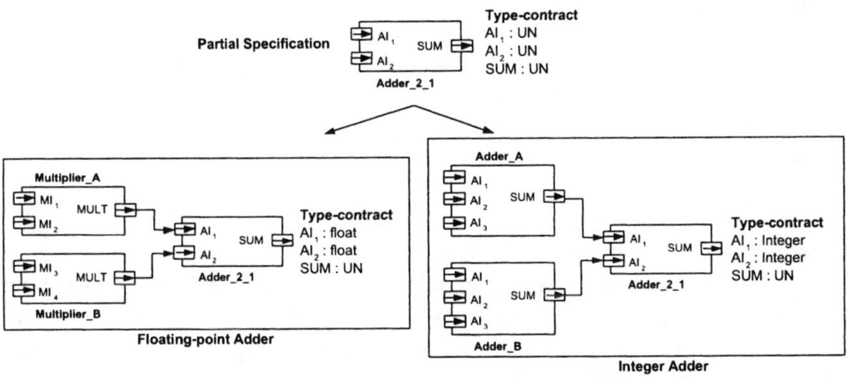

Figure 4.11 A generic Adder_2_1.

Another reason for allowing partially typed components/media is to avoid the tediousness associated with characterizing every component in the CCM. The types for some of these can be inferred based on their static-requirements or from their associations. For example, the CCMM provides media entities such as SPLITTER and MERGER for multiplexing and demultiplexing connections. Their input-output datatypes can be inferred from their static-requirements. Being able to infer the datatype of one port of the SPLITTER/MERGER instance requires the other ports to conform to it through type equivalence or subsumption.

Note that the specification of partially typed or untyped instances in the CCM may be such that the type inference fails to provide assignments that fully type the instances. This does not mean that a modeling error exists; rather sufficient type information is not provided by the modeler. This kind of incomplete specification is allowed through the CCMM. The advantage is that a polymorphic implementation is structural depicted using this capability of the CCMM. In C++ based HDLs such as SystemC, an IP block built using C++ template is polymorphic and has type variables as placeholders for concrete types, which are resolved during instantiation. Some examples of such implementations are buffers, FIFOs, memories, and switches. The modeler uses the CCMM for modeling a generic template of the system, into which IPs are substituted to construct various executable models. The automatic exploration of alternatives for the generic template is furthered through an incomplete CCM specification, because the untyped or partially typed instances are generic enough to match with a large number of implementations.

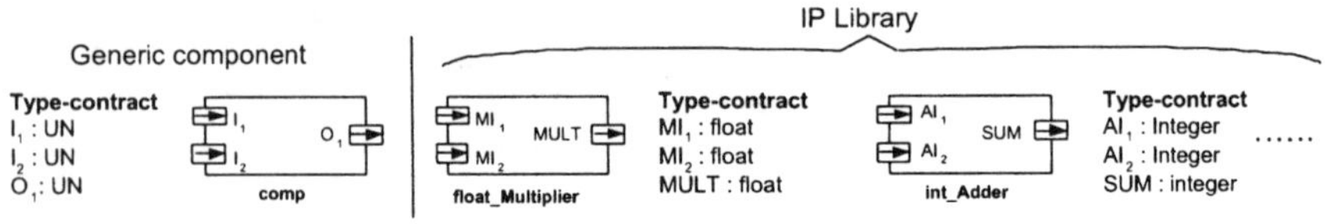

Figure 4.12 Instantiating a template from the IP library.

Consider the example of the generic component *comp* in Figure 4.12, which the modeler wishes to instantiate with an IP from the library. Two of the implementations from the IP library, namely *float_Multiplier* and *int_Adder* are also shown along with their reflected type-contracts.[1] Both of these can be used to instantiate the generic template of *comp* through a selection based on structural equivalence. However, the behaviors of these two IPs are different and one results in an incorrect substitution by the IComp after selection. This cannot be avoided unless the modeler provides more type-information for the interface of the component, since an untyped RTL component matches structurally with almost every IP with the same number of input and output ports. Hence, there is a trade-off between the rapid specification of the system template and the quality of the selected IPs for substitution.

Type Propagation

The type propagation engine begins by extracting the instances in the architectural template and populating data structures that are SETs. The extraction allows for an easy analysis and uniform treatment of similar instantiations. It also avoids multiple traversal of the model. A part of the extraction procedure is shown next.

Extraction Procedure

{Given a model CCM M, where LSET = SpSET = ChSET = . . . = ϕ}

Step 1 For each LEAF instance in M, populate LSET.

Step 2 For each SPLITTER instance in M, populate SpSET.

Step 3 For each CHANNEL instance in M, populate ChSET.

Step 4 . . .

1. The reflected type-contract of an implementation is simplified for illustration.

Step n For each instance from LSET, SpSET, . . .
  Step n.1 For each INPUTPORT object, populate IPSET.
  Step n.2 For each OUTPUTPORT object, populate OPSET.
Step m For each instance from ChSET, . . .
  Step m.1 For each READPORT object, populate RDSET.
  Step m.2 For each WRITEPORT object, populate WPSET.
    Step m.2.1 For each READPORT and WRITEPORT object, populate ARGSET.
    Step m.2.2 . . .

The engine applies inference techniques to statically derive datatypes for a partially typed instance based on its association- and static-requirement-constraints. Upon inference, the engine plugs in the inferred type and calls the type-checker to guarantee that the propagated type is consistent. The inference algorithms differ from entities to media. We will outline the algorithms for a component and a medium in this subsection. We also illustrate association- and static-requirement-constraint-based type inference.

Listing 4.9 Type Propagation on a LEAF instance.

```
// Algorithm for type propagation on a LEAF instance
// Based on association-constraint on IOPort objects

For each lf in LSET
  if(PartialTyped(lf) == true)
    For each op in lf.OPSET
      if(PartialTyped(op) == true)
       SIGSET = op.GetConnections("IOConnection","dst");
       For each sig in SIGSET
        ip = sig.FindDst();
        if(PartialTyped(ip) == false)
          // Datatype assignment
          op.dt = ip.dt; op.sz = ip.sz;
          break;
       TypeCheck(op); // Type-checking
```

Consider a LEAF instance from the LSET: the algorithm in Listing 4.9 tries to infer its input-output datatypes based on association-constraints. For a partially typed LEAF instance lf determined using function **PartialTyped**, the algorithm iterates over all partially typed output ports. For each such port (*op*), it extracts all the IOConnection instantiations with *op* serving as the source of the connection using function **GetConnections**. For each such connection, we obtain the destination (ip) using the **FindDst** function. If ip is fully typed then, *op* is assigned the *dt* and *sz* of *ip*. The function **TypeCheck** is used to verify the type assignment.

Listing 4.10 Type Propagation on a SPLITTER instance.

```
// Algorithm for type propagation on SPLITTER instance
// Association-constraint on IOPort objects and static-requirement-constraint on SPLITTER instance

For each sp in SpSET
if(PartialTyped(sp) == true)
  For each op_i in sp.OPSET //Part 1
  if(PartialTyped(op_i) == true)
    sig_i = op_i.GetConnections("SP2LConn","dst"); ip_i = sig_i.FindDst();
  if(PartialTyped(ip_i) == false)
    op_i.dt = ip_i.dt; op_i.sz = ip_i.sz; break;
  else
    For each op_j in sp.OPSET //Part 2
    if(op_i.portname <> op_j.portname && PartialTyped(op_j) == false)
      op_i.dt = op_j.dt; op_i.sz = op_j.sz; break;
    else
      if(op_i.portname <> op_j.portname && PartialTyped(op_j) == true)
        sig_j = op_j.GetConnections("SP2LConn","dst");
        ip_j = sig_j.FindDst();
          if(PartialTyped(ip_j) == false)
            op_i = ip_j.dt; op_i.sz = ip_j.sz; break;
          else
            For ip_k in sp.INSET //Part 3
            if(PartialTyped(ip_k) == false)
              op_i.dt = ip_k.dt; op_i.sz = ip_k.sz;
            else
              sig_k = ip_k.GetConnections("L2SPConn","src"); op_k = sig_k.FindSrc();
              if(PartialTyped(op_k) == false)
                op_i = op_k.dt; op_i.sz = op_k.sz; break;
  TypeChecked(sp);
```

Consider a SPLITTER instance (*sp*) from SpSET, the algorithm in Listing 4.10 tries to infer its input-output datatypes based on association and static-requirement-constraints. The algorithm tries to obtain a type assignment for each partially typed output port op_i of *sp*. It is divided into a three-part search. *Part 1* utilizes the association-constraint that allows the output port op_i of a SPLITTER instance to connect to an input port ip_i of a LEAF instance. This relation is enabled through the SP2LConn <<Connection>>. If *Part 1* turns up unsuccessful, then *Part 2* and *Part 3* exploit the static-requirement-constraint of *sp*. A splitter is a MUX, therefore all its multiplexing ports must have the same datatype (static-requirement of a SPLITTER). Therefore, *Part 2* searches for a typed output port op_j in *sp* to infer a datatype for op_i. If none of them are typed, then the search tries to obtain a type from the input ports (ip_j)s connected to these output ports (op_j)s (Similar to *Part 1*). If *Part 1* and *Part 2* fail, then *Part 3* tries to infer a type from the input port ip_k of sp. This also follows from the static-requirement-constraint (Similar to *Part 2*).

Similar algorithms are provided for all the entities in the CCMM. The type propagation fails to convert a partially typed instance into a fully typed entity on two accounts:

1. Type information is not sufficient to fully type the instance.
2. Type-checker detects an inconsistency during type propagation.

In the second case, the algorithm terminates by stating that the architectural template is not correct and there is a type inconsistency with respect to the specification of certain instances in the template. The modeler needs to go back to the architectural template and rework the type-contract to remove the conflicting scenario.

4.2.3 XML Translation

The translator generates an intermediate representation for the CCM using XML. This XML representation (XML-IR) is validated against a document type definition (DTD) for wellformedness. The translator uses a templated code generator that when given a visual component or media prints out the corresponding XML fragment. The XML-IR serves as input to the introspective composer. Note that instead of using our XML-IR, the SPIRIT [12] XML format can be used for exporting the model.

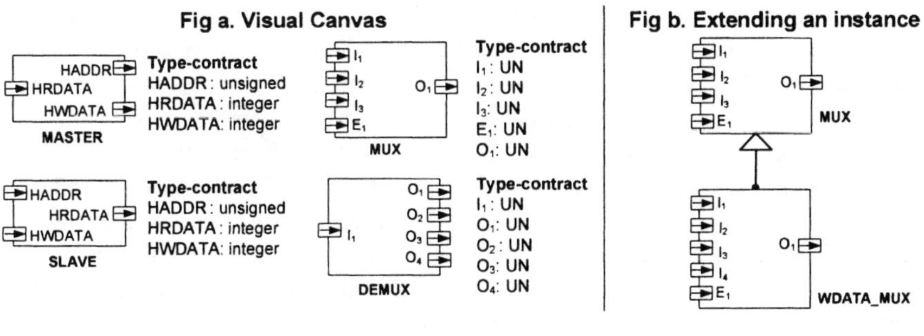

Figure 4.13 (a, b) Visual canvas and subtype instance.

4.3 CASE STUDIES

We discuss the CCM created for the AMBA AHB bus [13] and the simple bus from the SystemC distribution [14].

4.3.1 AMBA AHB RTL Bus Model

The AMBA AHB has three master components with identical interface, four slave components with identical interface, an arbiter, and a decoder as shown in Figure 2.2. The master components communicate with the slave components through MUXes, DEMUXes, and a set of signals. The multiplexing is driven by the arbiter and the decoder. In order to quickly model the architectural template of the AMBA bus using our CCL, the reusable modules in the system are identified. The reusable components and media are created on the visual canvas (part of the modeling framework), which are instantiated through a *copy-paste* action and characterized to create the CCM for the system. For the description of the AMBA CCM, we create a master, slave, generic MUX, and DEMUX, which are placed on the canvas to facilitate reuse.

Visual Canvas

The master is modeled as a LEAF instance with three IO ports, namely **HADDR**, **HRDATA**, and **HWDATA**. The **HRDATA** is an input port receiving data of type `int`, whereas **HWDATA** is an output port similar to **HRDATA**. The **HADDR** is a 32-bit `sc_uint` output port. These type-bindings are used to characterize the master's interface, thereby producing a fully typed component (type-contract is complete). This is possible for the AMBA CCM, because the master components are identical. However, in general, the components and media on the canvas are modeled generically, which provide more flexibility in terms of reuse. The slave component on the canvas is modeled similar to the master with three IO ports, namely **HADDR**, **HRDATA** and **HWDATA**, where **HADDR**, and **HWDATA** are input ports and **HRDATA** is an output. A fully specified type-contract is also provided for the slave component that is similar to the master component's type-contract. On the other hand, the MUX and DEMUX on the canvas are modeled generically, because their type-contracts differ based on the (de-)multiplexed signal. The MUX is modeled as a MERGER instance with three input ports (I_1, I_2, and I_3) and an output port (O_1) for multiplexing, which structurally depict the requests from the three masters that are processed one at a time. The MUX also has an external input port (E_1) for the driver that selects the signal to be multiplexed. The DEMUX is modeled as a SPLITTER instance with one input port (I_1) and four output ports for demultiplexing (O_1, O_2, O_3, and O_4), which describe the read-write accesses being demultiplexed to one of the four slaves components. The visual canvas with the components and media that are used for the CCM description of the AMBA AHB is shown in Figure 4.13(a).

AMBA CCM

The CCM of the AMBA AHB bus is constructed in the following manner: The master component from the visual canvas is instantiated three times and the name characterization **Master_1**, **Master_2**, and **Master_3** are provided for the three instances. Similarly, the slave component is instantiated four times and characterized. We create new components, namely **Arbiter** and **Decoder** to model the arbiter and decoder in the AMBA AHB, since none of the components from the canvas can be reused for this purpose. Finally to compose these components, we need to model the signals by instantiating the MUX and DEMUX from the canvas. To multiplex the **HADDR** of the three masters, the MUX is instantiated and characterized as **ADDR_MUX**. Connecting **Master_i**'s **HADDR** port to the **ADDR_MUX** port (I_i) activates a constraint on I_i as well as on all the other multiplexing ports. The port I_i is automatically renamed as **HADDR_Master_i** through the consistency checker, which is also part of the characterization. For example, creating a connection such as c_1(**MASTER_1.HADDR**, **ADDR_MUX**.I_1) leads to the following automatic characterization through the type propagation engine:

$$f_{id}(I_1, \textbf{HADDR\_Master\_1}) \text{ and } f_{dt}(\textbf{HADDR\_Master\_1}, \texttt{sc\_uint})$$
$$f_{dt}(I_2, \texttt{sc\_uint}), f_{dt}(I_3, \texttt{sc\_uint}), \text{ and } f_{dt}(O_1, \texttt{sc\_uint})$$
$$f_{sz}(I_1, 32), f_{sz}(I_2, 32), f_{sz}(I_3, 32), \text{ and } f_{sz}(O_1, 32)$$

The connection not only binds the associated input ports, but also the other multiplexing ports due to the static-requirement of a MERGER that requires each of its instance to have type equivalent or subsumed ports for multiplexing. While connecting the **HADDR** port of the other masters, the type-bindings created from **Master_1** will enforce a type check. In this case study, the masters have identical interfaces, therefore this check will not be violated. However during the connection of components with distinct boundaries, the characterization of a MERGER instance from a single component can be used to detect type inconsistencies among the other components communicating through it. These inconsistencies are detected by the type-checker.

The external input of **ADDR_MUX** is connected to the **Arbiter** and characterized accordingly. Another MUX is instantiated to multiplex the **HWDATA** port of the masters. This instance is named **WDATA_MUX** and is similar to the **ADDR_MUX** with the difference that the type-binding propagated is `int`. This points out the need for a generic MUX that could be reused, as opposed to having a fully typed MUX in the canvas.

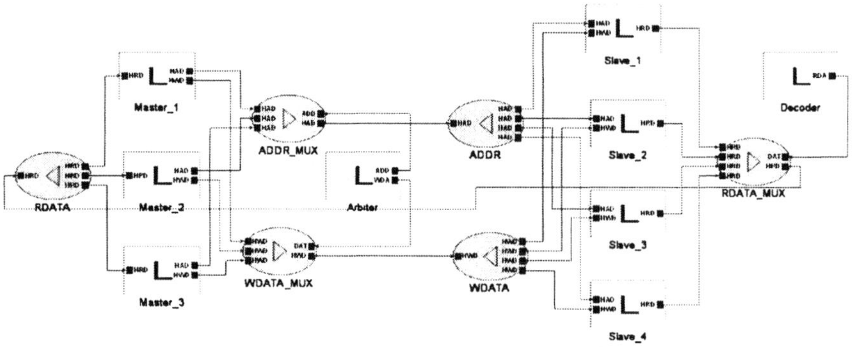

Figure 4.14 AMBA AHB CCM.

Besides promoting a reuse-based modeling flow of fixed blocks, the visual canvas also allows reuse through flexible blocks. A fixed block is an instance of the component/medium described on the canvas, which is reused and characterized in the CCM without changing its internals. On the other hand, a flexible block is an instance that is reused in the CCM after modifying its internals. The modification is limited to addition of objects and relations to the instance, but deletion of existing objects/relations within the instance is not allowed. This allows the modified instance (child) to be a *subtype* of the instance being modified (parent). Therefore the OCL constraints on the parent are propagated onto the child, which enforces its consistent usage in the CCM. In the AMBA bus model, consider the **HRDATA** input port of a master. This input to the master is obtained by multiplexing the output of the four slaves based on the decoder's input that addresses the master whose request is being processed. In order to model the **HRDATA** multiplexing, the MUX instance from the canvas is reused **(WDATA_MUX)**, after extending it to handle four inputs as shown in Figure 4.13(b). Furthermore, connecting the **WDATA_MUX** ports to the slaves ports (O_1, O_2, O_3, and O_4) and the port of the **Decoder** derives its type-contract through inference. In this case, the parent (MUX) is generic, therefore the type-bindings of the subtype are consistent. Similar to the multiplexing, the DEMUX is instantiated and characterized to create the complete ABMA CCM as shown in Figure 4.14.

4.3.2 Simple Bus TL Model

The TL model of the simple bus in [14] consists of three distinct masters that interact with two slaves through a bus and an arbiter as shown in Figure 2.3. The masters describe a blocking, nonblocking, and direct protocol to communicate with the slaves. The read-write requests from the masters are placed on the bus and prioritized by the arbiter. The slaves provide access to the memory, where one of the slaves provide a fast access and the other a slow clocked access. The three masters, two slaves, and the arbiter have distinct transaction interfaces, which were described using our CCL as shown in Figure 4.15. The trapezoid with a centered I is the INTERFACE construct in our CCL that is used to describe a transaction interface, similar to an `sc_interface` construct in SystemC.

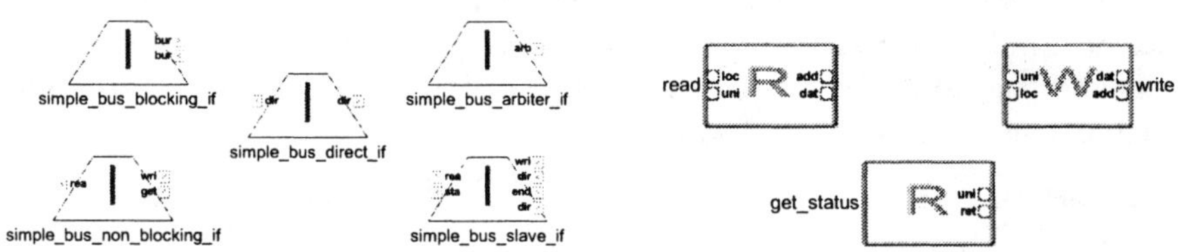

(a) Modeling the interfaces. (b) The **simple_bus_non_blocking_if**.

Figure 4.15 (a, b) Simple bus transaction interfaces.

Visual Canvas

The masters have distinct protocols and the only structural similarity is that they are clocked. Therefore, the reusable component for a master is an instance of the LEAF entity with a clock port. The slaves on the other hand have identical boundaries with the exception that the slow slave has a clock port, which is not made part of the reusable description. The reusable slave is modeled as a CHANNEL instance as shown in Figure 4.16(a), which implements the **simple_bus_slave_if** interface as shown in Figure 4.16(b).

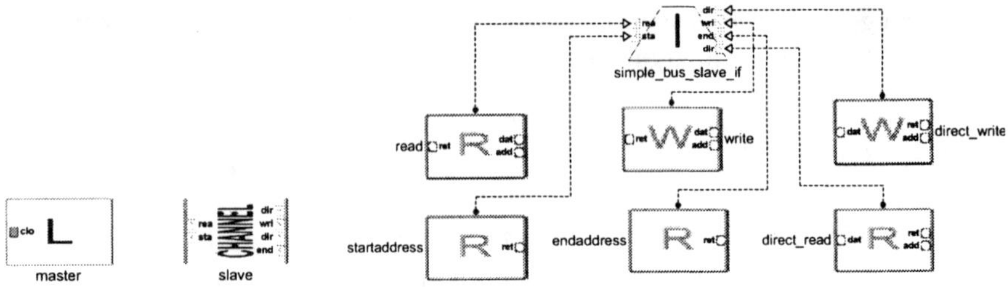

(a) Component and media. (b) The **slave** media.

Figure 4.16 (a, b) Visual canvas.

Simple Bus CCM

The CCM of the simple bus is constructed in the following manner: The master component from the canvas is instantiated and extended to include the corresponding ports to use the blocking, nonblocking, or direct interface. For example, the **simple_bus_master_non_blocking** component obtained by instantiating the **master** from the canvas is extended with read and write ports to use the **simple_bus_non_blocking_if** interface as shown next:

`//` *`read`* `function of the` **`simple_bus_non_blocking_if`**
f_{id}(READPORT, read)
f_{id}(read.Argument, prior), f_{id}(read.Argument, data), f_{id}(read.Argument, addr), f_{id}(read.Argument, lock)
f_{dt}(prior, `int`) and f_{iotype}(prior, output),
f_{dt}(addr, `int`) and f_{iotype}(addr, output),
f_{dt}(lock, `bool`) and f_{iotype}(lock, output),
f_{dt}(data, `int`) and f_{iotype}(data, input)
`//` *`write`* `function of the` **`simple_bus_non_blocking_if`**
f_{id}(WRITEPORT, write)
f_{id}(write.Argument, prior), f_{id}(write.Argument, data), f_{id}(write.Argument, addr), f_{id}(write.Argument, lock)
f_{dt}(prior, `int`) and f_{iotype}(prior, output),
f_{dt}(addr, `int`) and f_{iotype}(addr, output),
f_{dt}(lock, `bool`) and f_{iotype}(lock, output),
f_{dt}(data, `int`) and f_{iotype}(data, output)
`//` *`get_status`* `function of the` **`simple_bus_non_blocking_if`**
f_{id}(READPORT, get_status)
f_{id}(get_status.Argument, prior),

f_{id}(get_status.Argument, return_type)

  f_{dt}(prior, `int`) and f_{iotype}(prior, output),

  f_{dt}(return_type, `string`)

Similarly, the **simple_bus_master_blocking** and **simple_bus_master_direct** components are created to describe the other masters. These components make read and write requests to access the slave memory through the bus (**simple_bus**). The **simple_bus** media is a BUS instance that implements the **simple_bus_blocking_if**, **simple_bus_non_blocking_if**, and **simple_bus_direct_if** interfaces as shown in Figure 4.17.

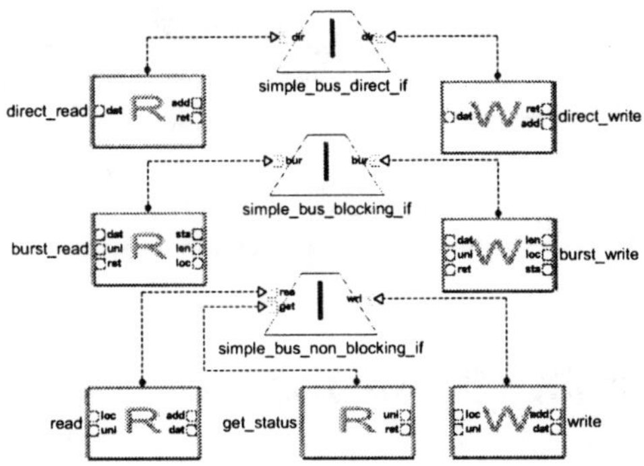

Figure 4.17 The **simple_bus** media.

The arbiter (**simple_bus_arbiter**) is described as a CHANNEL instance, which implements the **simple_bus_arbiter_if** interface. The slaves are described using the **slave** media from the canvas (shown in Figure 4.16). In order to describe the fast slave, the **slave** media is used as a fixed block and characterized into **simple_bus_fast_mem**, whereas to describe the slow slave, the **slave** media is extended into **simple_bus_slow_mem** with an additional port for the clock. Note that the reason to describe the slaves and the arbiter as CHANNEL instances is to create a CCM very close to the implementation of the simple bus in the SystemC distribution. The CCM of the simple bus is obtained by connecting the masters, slaves, and the arbiter to the bus through their respective interface ports. For example, connecting the **read**, **write**, **get_status** ports of the **simple_bus_master_non_blocking** component to the concrete ports (**read**, **write**, and **get_status**) in the **simple_bus** media allows the master to make requests on the bus. This connection translates to an `sc_port` in the SystemC implementation of **simple_bus_master_non_blocking** component that is composed through the non_blocking interface. These connections are only allowed if the ports match with respect to function signatures. The CCM created for the simple bus is shown in Figure 4.18.

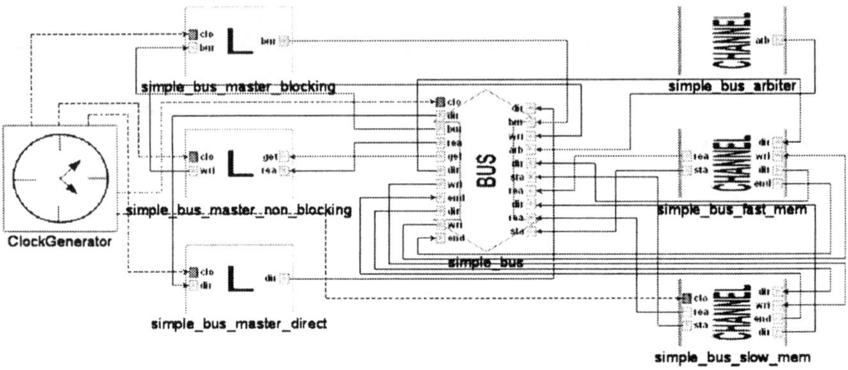

Figure 4.18 Simple bus CCM.

4.4 DESIGN EXPERIENCE AND SUMMARY

Our experience from using the modeling framework in MCF is summarized as follows: (1) it is a drag-and-drop environment with minimal effort in modeling systems, (2) it allows for rapid model development through a visual canvas of reusable elements, (3) it provides for modeling time checks to enforce certain correctness constraints, (4) it is easy to model at different abstractions as well as across abstractions, (5) it provides a natural way to describe hierarchical systems, and (6) it is highly extensible and customizable for future requirements, since it is built using the metamodeling language of GME.

To assess the scalability and usability of the visual modeling framework of MCF, we have built a wide variety of models. These are examples from the SystemC distribution and designs obtained from our industrial contacts such as INTEL. Some of the reasonable sized designs modeled are a microarchitecture of a RISC CPU with MMX capability [14] and a digital equalizer system [15], which are downloadable from [10]. When dealing with complex designs, scalability of a visual modeling tool is a concern. Therefore MCF provides two modeling modes: (1) visual, based on the CCL and (2) textual, based on the TCCL. Having a textual language addresses the scalability issue with the visual modeling tool and allows the modeler to take advantage of checkers and the other analysis engines, which are a part of MCF. One disadvantage of taking the textual modeling route is the suspension of modeling time constraint checks. The application of these checks is postponed until the model is completed and imported into the modeling framework. If the modeler has made numerous modeling errors, then the checks could be overwhelming. This complexity is surmounted by performing an incremental modeling, where parts of the template are built and imported into the framework to perform the constraint checks.

Final Goal

An MCF modeler creates an abstract architectural specification of the system that is independent of any IP implementation specifics and then through a few user initiated actions on the GUI and creates executable models. The task of the MCF besides enabling a specification is to take the architectural template of the system and automatically plug in matching IPs from a library through type inference. The type inference facilitates automatic selection of matching IPs for the abstract components in the architectural template and composes them to create simulation models for the design alternatives. The advantage is the reduction in design time and tediousness associated with constructing high-level models for DSE through the automated tool support.

References

[1] A. Ledeczi, M. Maroti, A. Bakay, G. Karsai, J. Garrett, C. Thomason, G. Nordstrom, J. Sprinkle, and P. Volgyesi, "The Generic Modeling Environment," *Proceedings of Workshop on Intelligent Signal Processing* Budapest, Hungary, May 2001.

[2] Object Management Group, "Unified Modeling Language," http://www.uml.org/.

[3] J. Warmer and A. Kleppe, *The Object Constraint Language: Getting Your Models Ready for MDA*, 2nd Edition, Reading, MA: Addison-Wesley, 2005.

[4] B. DuCharme, *XML: The Annotated Specification*, Upper Saddle River, NJ: Prentice Hall, 1998.

[5] A. Jantsch and H. Tenhunen, *Networks on Chip*, San Francisco, CA: Morgan Kaufmann Publishers, 2003.

[6] F. Ghenassia, *Transaction-Level Modeling with SystemC: TLM Concepts and Applications for Embedded Systems*, New York: Springer-Verlag, 2006.

[7] VHDL, http://www.vhdl.org/.

[8] Verilog, http://www.verilog.com/.

[9] System Verilog, http://www.systemverilog.org/.

[10] D. A. Mathaikutty and S. K. Shukla, "A Metamodel for Component Composition," FERMAT Group, Virginia Tech, Tech. Rep. 2005-13, December 2005.

[11] Y. Xiong, "An Extensible Type System for Component Based Design," University of California Berkeley, Electrical Engineering and Computer Sciences, Ph.D. Thesis, 2002.

[12] Spirit Consortium, "SPIRIT Schema Working Group," http://www.spiritconsortium.org/.

[13] ARM, "AMBA Specification 2.0," http://www.arm.com, 1999.

[14] OSCI Group, SystemC Web site, http://www.systemc.org/.

[15] FERMAT Group, MCF Web site, http://fermat.ece.vt.edu/MCF/MCF.htm, 2005.

Chapter 5

IP Reflection and Selection

MCF employs introspection metadata reflected from SystemC IPs to analyze and select implementations that can be used to create an executable for the abstract specification. The metadata that would facilitate IP selection and composition includes register-transfer level (RTL) ports and datatypes, transaction-level (TL) interface signatures, hierarchy information, polymorphic characterization, and so forth.

In order to reflect the metadata from IPs, the important questions that need to be answered are: "What is the metadata of interest?" and "How accessible is it?" The first question would depend on how the extracted information will be used. The second question depends on how the metadata is specified. If the information is given through annotations to the implementation such as comments and pragmas, then the reflective mechanism would easily extract these. In most cases, the information of interest is not available in a straightforward manner and therefore extraction techniques would require mining the design to infer the metadata. This problem is more prominent with languages such as C++, because of their high-level expressiveness as well as their capability to enable type indirections (through constructs such as `typedef` in C++) and allow for generic types (polymorphism through C++ templates). This problem is addressed through this chapter, where we show some of the challenges involved in mining metadata from designs written in a highly expressible language such as C++. Our work mainly focuses on IPs specified in SystemC and the metadata of interest pertains to structural types of the IP constituents and their communication, which is utilized by MCF for IP selection.

5.1 METADATA FOR IP COMPOSITION

In this section, we address the question: "What is the necessary composition-related metadata?" We state the relevant IP-level metadata independent of SystemC and then discuss how this information is specified through a SystemC description. An IP is a specific implementation of a set of standard blocks and their interaction in the form of electronic design or silicon. Please note that our work only is relevant to soft-IPs. These standard blocks are called components and their interaction is dictated by the underlying communication model. The metadata on a component basically describes its structural type. It defines the datatype binding of the different structural aspects of a component. The structural aspects of a component are closely tied to the abstraction level of its specification. For example, a component specified at the RTL abstraction is structurally depicted through ports, parameters, and hierarchical instantiations. The datatypes specified on these structural aspects are captured as the contract for the component and employed during selection and composition of IPs. In terms of the communication model, components communicate through different means such as signals, buses, channels, and switches. Therefore, we associate a structural type with these communication primitives, which also need to be extracted as metadata. The reason for separating components from their communication model is to have a one-to-one correspondence with the architectural template during the selection process. The architectural template segregates components and media; therefore we distinguish between metadata on components and communication primitives in an IP. Some of the other relevant metadata are more to enable specific styles of selection. They may not be explicitly specified in the IP by the implementer, but are inserted by the library engineer in order to add the IP to the library and ready it for selection.

In a SystemC description, some of the metadata are available in a straightforward manner, whereas some others require data mining techniques in order to identify them. The abstraction levels supported through SystemC are RTL and TL. Therefore, the metadata is classified into information related to components and channels. This is because the communication models supported are either signal-based for RTL or channel-based for TL. The channel is implemented such that it simulates the required communication model, be it a bus or a switch. The metadata on an RTL component contains the set of parameters and input-output ports. For these, information related to the datatype and bitwidth are essential constituents of the metadata. Therefore it is important to correctly extract these through different mining techniques. For a component/channel at TL, it contains a set of interface ports, and interface access methods. Therefore the metadata would contain the function signatures that embed the function arguments and return-types. For these arguments and return-types, the datatype and bitwidth need to be correctly extracted. The hierarchical structure of a component and the types of the hierarchical instances are also essential metadata. All of these are automatically extracted and represented through our methodology discussed in Section 5.3. Note that we do not extract any timing information; therefore our TL is more of the programmer's view of abstraction.

Furthermore, to facilitate the addition of polymorphic IPs to our IP library, we allow extra information to be annotated to the IP. This is not necessary, but helpful to avoid incorrect selections. Consider the following scenario, where a polymorphic adder **A**, which internally performs only integer additions, is provided for library inclusion. If no constraints are provided on the polymorphic aspect of **A** that restricts it to integer datatypes, then there is a possibility that during selection of a floating point adder, **A** might end up being selected. Therefore, we allow the library engineer to provide the type constraints as annotations. A better approach is to extract the behavioral type of the adder, but currently our approach is limited to structural types. Some of the other metadata that can be annotated are used in cataloging IPs, such as naming format, and version numbers. Note that these are not necessary, but are provided to enable a well-defined selection process. Furthermore, many times the type information is provided through different indirections. These arise due to the expressiveness of C++. Therefore, the metadata needs to be identified and extracted correctly, which is a nontrivial task. We provide some discussion on the kind of indirections that cause problems during metadata extraction and reflection and our solution approach to these problems.

5.2 METADATA ON A SYSTEMC IP SPECIFICATION

SystemC is a library of C++ classes and every design is an instantiation of some of these classes. Therefore, class-based extraction is the primary task during metadata mining. The extracted classes help in identifying the different modules, interfaces, and further which of these modules describe components and which implement interfaces to describe channels, as shown in Figure 5.1. In order to find this kind of information, we extract every class and the associated inheritance tree from the SystemC model.

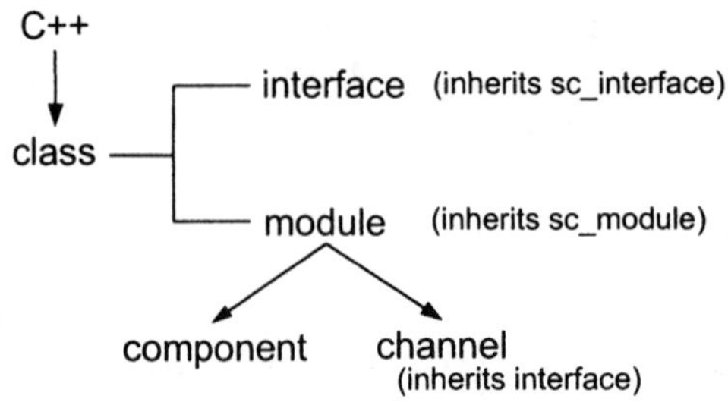

Figure 5.1 Extraction of classes.

```
// Description Styles
SC_MODULE(X) { ... };

class X : public sc_module { ... };
```

Figure 5.2 Module description styles in SystemC.

Module Identification

A module is identified in a straightforward manner, either it is an *sc_module* (base_module) or a class that publicly inherits from the *sc_module* (derived_module). In Figure 5.2, we illustrate two common structural description styles for a module X. We search the associated inheritance tree looking for an **sc_module** and if one is encountered, then that class gets tagged as a module.

Interface Identification

An *interface* in SystemC fulfills the following requirement:

- Must be an abstract base class (in C++);
- Inherits from **sc_interface**;
- Specifies a set of access methods for a channel, but provides no implementation.

Therefore, identifying *first-level interfaces* requires searching the inheritance tree for an **sc_interface** and tagging them as interfaces. It is common to have interfaces that inherit from first-level interfaces that we call *next-level interfaces*. This type of interface inheritance can span to any depth and therefore cannot be extracted in a single pass of the model. We implement an interface extraction procedure that recursively identifies and extracts the metadata on an interface irrespective of the inheritance hierarchy.

Component Identification

The extraction of metadata from a module class is very different from an interface class, as their internals differ. A module contains parameters, ports, and processes, whereas the interface declares function signatures. Modules are categorized into computational components and communicative channels. A *component* describes a behavior through its input-output assignments, whose implementation in SystemC is provided through classes. It can be identified by searching for an **sc_behavior** as shown in Figure 5.3.

```
class CompA : public sc_behavior { ... };
```

Figure 5.3 Component description in SystemC.

Channel Identification

A *channel* serves the purpose of a communication medium and models the interaction between components. In SystemC, a channel fulfills the following requirements:

- Must derive from **sc_channel**;
- Must derive from one (or more) classes derived from **sc_interface**;
- Provide implementations for all pure virtual functions defined in its parent interfaces.

The first requirement helps in structurally identifying a channel, which boils down to searching for an **sc_channel** in the inheritance tree of the class. The second requirement helps in figuring out the list of interfaces this channel implements. The last requirement is very important as it helps in extracting the access methods of the identified channel, which is essentially the metadata captured for a channel.

```
// Interface description
class write_if : public sc_interface {};
class read_if  : public sc_interface {};

//Channel description
class fifo : public sc_channel,
             public write_if,
             public read_if
{ ... };
```

Figure 5.4 Interface and channel description in SystemC.

As illustrated in Figure 5.4, we have two interfaces, namely *read_if* and *write_if* and a channel *fifo* that implements these interfaces. The reason for categorizing modules into components and channels is shown in Figure 5.5, which allows us to structurally distinguish the different modules.

```
// Channels
typedef sc_module sc_channel;
//Components
typedef sc_module sc_behavior;
```

Figure 5.5 Module classification.

If the above requirements of a channel are not fulfilled, then we need another way to distinguish a channel from a component. It is a common practice to inherit from the sc_module to describe both components and channels. In this case, we structurally distinguish the two using the following requirement:

- Requirement 1: A channel derives from one (or more) class(es) derived from sc_interface and a component does not.

Another reason for categorizing modules is that the internals of a channel and a component differ and mandate the parser to perform different extractions.

Metadata on Interfaces

In order to avoid problems with multiple-inheritance and provide safe interfaces in SystemC, interfaces never implement methods or contain data members. Given the way of describing an interface in SystemC, the only metadata of interest is function signatures of access methods as shown in Figure 5.6. To obtain a function signature, the parser extracts the return-type and function arguments of the method in terms of their name, datatype, and bitwidth.

```
//1st level interface
class simple_bus_blocking_if : public sc_interface
{
// Interface access method
virtual void read(unsigned int unique_priority,
int *data, unsigned int address, bool lock) = 0;
...
}
```

Figure 5.6 Interface access method *read* in SystemC.

Metadata on Components

A component in SystemC is a collection of parameters, ports, and processes. Ports play a crucial role in describing the structure of a component, whereas processes describe the computation. Therefore, the main task while mining metadata of a component is to correctly extract its parameters, ports, and port level details. The ports can be classified into two types based on the abstraction level of the component. A component could be described at the RTL abstraction, which requires it to have input-output ports. It can also be described at the transaction level, where it communicates through interface-level ports.

```
// Input port
sc_in<sc_int<16> > in_real;
// Interface port
sc_port<read_if> data_in;
```

Figure 5.7 Ports in SystemC.

A component's port level RTL implementation is provided through the constructs **sc_in**, **sc_out**, and **sc_inout** in SystemC as shown in Figure 5.7. A component's interface-level TL implementation is provided through the **sc_port** construct in SystemC, which gives the component access to the methods of the interface. Similar to the input-output ports, we extract the clock port, which is provided using the **sc_in_clk** and **sc_out_clk** constructs in SystemC.

SystemC allows for structural hierarchy by instantiating modules within a module. The metadata on the hierarchical structure of a component boils down to its module hierarchy.

Metadata on Channels

The main purpose of channels in SystemC is to facilitate TL modeling, where the communication semantics are rendered through the interface-level function calls. As a result while mining for metadata on a channel, we extract its parameters, interface-level function declarations, interface ports, and clock ports to describe its structure. Looking at the interfaces that a channel inherits, we can separate the interface access methods from all the function declarations.

It is commonly seen as an industry practice to avoid the usage of sc_port construct during TL modeling (TLM) due to issues related to synthesis. However, to facilitate TLM, it is common to have a component inherit from the channel implementing the interface. This would violate our first requirement that makes it possible to distinguish a component from a channel. Therefore, we introduce an alternate requirement.

- Requirement 2: If a component derives from one or more classes (besides the **sc_module**), then these classes should be channels.

Metadata on Polymorphic Components and Channels

In C++ based HDLs such as SystemC, a component/channel described using a partially specified C++ template is polymorphic and uses type variables as placeholders for concrete types that are resolved during instantiation. Some examples of such implementation are buffers, FIFOs (Figure 5.8), memories, and switches.

```
//Polymorphic component description
template < class DataT, unsigned int size >
class fifo : public sc_module { ... };
```

Figure 5.8 Generic FIFO.

In order to capture metadata on these polymorphic descriptions, we start by extracting the template expression. Then for the implementation, the polymorphic parts are the ports or interface functions. Examples of such an I/O and interface port are shown in Figure 5.9. Therefore, we need to extract the polymorphic nature of these as a part of the metadata of ports.

```
//Polymorphic I/O port
sc_in<DataT> data_in;
//Polymorphic interface port
sc_port<read_if<DataT> > data_in;
```

Figure 5.9 Polymorphic ports.

For an interface function, the polymorphic nature can be attributed to function arguments or return-type (Figure 5.10) and is dealt within a similar manner as polymorphic ports.

```
//Polymorphic function argument
void read(DataT &);
//Polymorphic return-type
DataT & write();
```

Figure 5.10 Polymorphic interface functions.

Another place where the effect of templates needs to be taken into account is during hierarchical embedding. If the submodules of a module are templated, then during the extraction of the metadata on structural hierarchy, the polymorphic nature of submodules should be captured.

Metadata from Annotations

The designer can explicitly annotate the code with metadata and it is commonly carried out through comments or pragmas. We allow the designer to insert comment-level annotations into the implementation, which would guide the constraining aspect of the polymorphic component/channel as well as the selection process. The user inserts hints through comments of a certain format into the code from which we obtain the metadata. One example of where we employ annotations is to constrain a generic implementation. A comment that specifies a type-constraint on some component/channel has the format shown here:

constraint : *name* $< arg_1, \ldots, arg_n >$, where $arg_i = type_1; type_2; \ldots$

The comment-level constraint has a format similar to the template expression. The *name* describes the component on which the constraint is enforced. arg_i has a one-to-one correspondence to a template argument based on its position in the template expression. However, arg_i captures legal C++ or SystemC types that could be instantiated instead of the placeholders in the template expression. The library engineers insert the specialized comments into the relevant IPs. These are extracted as a part of the metadata on the component/channel on which the constraints are specified. Similarly, selection hints can also be provided as annotations in the code.

Mining Metadata

We extract fragments of the implementation, which are language-level specifics that are processed to guide the extraction of certain metadata that are not available in a straightforward manner. The language-level specifics are indirections that ease the designer's implementation process. C++ provides many high-level constructs (indirections) to eliminate the repetition or tediousness associated with the usage of lower-level constructs. These indirections hide metadata from a straightforward extraction from the implementation. Therefore, we extract these language-level specifics and process these to reveal the hidden metadata.

An example in C++ is name aliasing introduced through constructs such as *typedef*. The usage is to insert a simple alias for a long statement that is used in multiple places in the implementation. We call the usage of these constructs *type indirection*, since they hide the actual type of a port or interface during extraction. As a result, we perform a search and replace the alias with the actual legal type. This requires identifying a name alias, the actual type associated with the alias, and then updating it with the actual type. Therefore, we extract the usage of the following access modifiers *#defines*, *typedef*, and *const*. The need to extract these constructs such that they can be processed in search of any indirection is not trivial. We achieve this by implementing a replacement procedure that incrementally replaces the *type indirection* to arrive at the actual datatype.

In Figure 5.11, we illustrate the result obtained after performing reflection on the RTL FIR Filter model from the SystemC distribution [1]. The port level specifics and signal connectivity extracted as a part of the metadata is also shown.

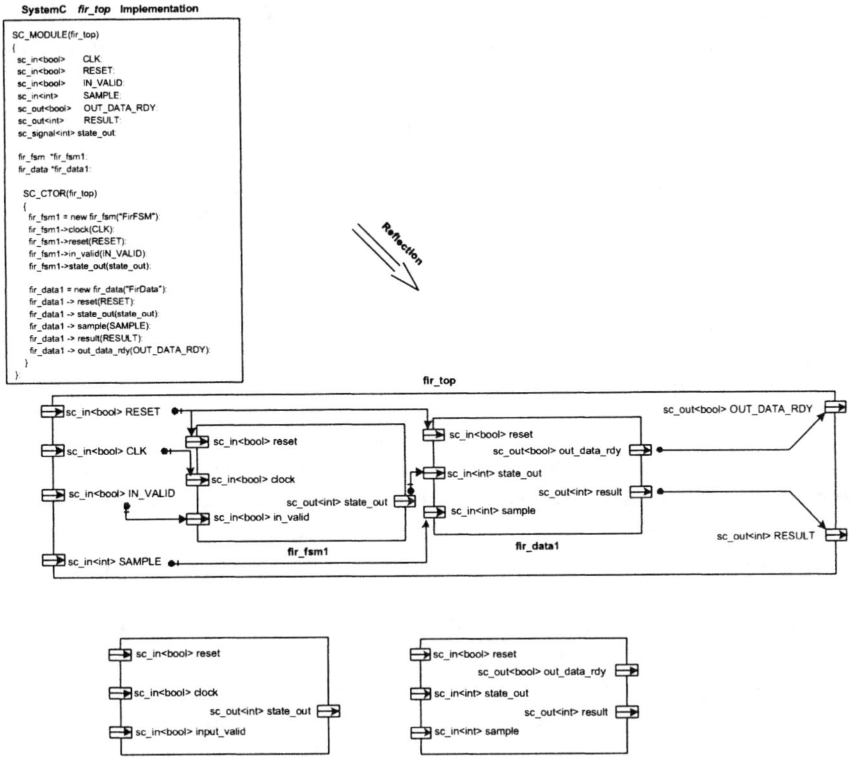

Figure 5.11 Structural reflection of FIR filter.

5.3 TOOLS AND METHODOLOGY

Given a library of SystemC IPs, we achieve automated extraction of metadata from these using tools such as KaSCPar and Xerces-C++ parsers and a methodology built on top of two languages, namely C++ and XML. The objective is to facilitate IP selection and composition by the CCF employing the extracted metadata from the IP library. The methodology has different stages that extract, infer, and constraint the metadata from the given library of IPs. The design flow for our methodology is shown in Figure 5.12. It has three stages, which begins with a library of SystemC designs and ends with an XML document object model that embeds the metadata of these designs. It is called the component DOM (cDOM) and serves as the primary input to the CCF. In the following subsections, we elaborate on the various stages.

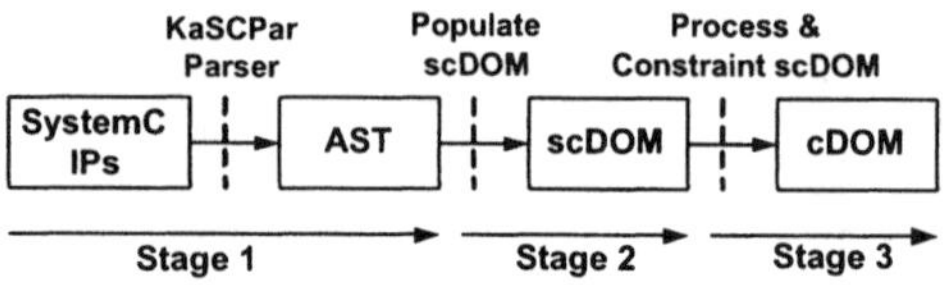

Figure 5.12 Design flow of the methodology.

5.3.1 Stage 1: SystemC Parsing

The input to this stage is a design written in SystemC and compiled using a C++ compiler. We use the KaSCPar SystemC parser to traverse the designs and print out the corresponding XML, which is an AST. It contains some metainformation like column number, line number, and file name as a reference to the original SystemC file and is saved as an attribute of each AST token. The comments of source files are also saved as special tokens in the AST.

In this stage, we obtain a more generic C++ based extraction than a specific SystemC metadata extraction that can be geared for CCF usage. As a result, the AST is too detailed and contains irrelevant information. Furthermore, to reflect certain metadata of a design, we need to process the first-level extraction and perform analysis. This analysis tries to decipher the indirection introduced by high-level constructs of the language. Therefore, all further processing is based on the AST.

5.3.2 Stage 2: AST Parsing and DOM Population

In this stage, we parse the AST using the Xerces-C++ parser [2] to extract a subset of metainformation necessary for IP composition and populate the internal data structure of the parser. The data structure is a Document Object Model (DOM) that serves as a platform- and language-neutral interface that will allow programs and scripts to dynamically access and update the content of the DOM. Therefore, attaching the DOM with an introspective architecture that implements the querying interface can be achieved with ease. The output of this stage is called the **sc_DOM**, which is given as input to the last stage. We briefly describe the subset of the AST extracted to populate the **sc_DOM** below.

The primary task as explained in the previous section is the extraction of classes, which are categorized into components, channels, and interfaces. The class information is extracted from the AST and used to populate the **sc_DOM** through a **class_specifier** token. This token has an internal structure that captures the data members and member functions associated with the class. It also captures the inheritance tree of this class, which is essential for the categorization of the SystemC modules. We illustrate some of the tokens and the metadata captured during the population of the **sc_DOM**. In this section, we address the question, "How do we represent the relevant metadata?"

Interfaces

The metadata captured for an interface is the function signatures. For the simple_bus_blocking_if read access method shown in Figure 5.6, the metadata is captured by a **function_declaration** token shown in Figure 5.13 and wrapped in an **interface** token.

```
<interface name="simple_bus_blocking_if">
    <function_declaration name="read"
    partial_spec="false" return_type="void">
        <argument datatype="unsigned int"
        name="unique_priority" partial_spec="false"/>
        <argument datatype="int" name="data"
        partial_spec="false" ptr="true"/>
        <argument datatype="unsigned int"
        name="address" partial_spec="false"/>
        <argument datatype="bool" name="lock"
        partial_spec="false"/>
    </function_declaration>
</interface>
```

Figure 5.13 The **function_declaration** token *read* in XML.

Components

The metadata captured for a component is the port level specifics. The parser generates a **port_declaration** token for each port encountered in the component, which embeds the name, datatype, and bitwidth of the port in its attributes. The extraction of an interface port for a component at TL is performed very similar except that the datatype attribute of the **port_declaration** captures the interface name. In Figure 5.14, we illustrate the **port_declaration** token for the input port *in_real* and interface port *data_in* of Figure 5.7.

```
// Input port
<ports_declaration IOtype="sc_in" bitwidth="16"
datatype="sc_int" name="in_real" partial_spec="false"/>
// Interface port
<ports_declaration IOtype="sc_port" datatype="read_if"
name="data_in" partial_spec="false"/>
```

Figure 5.14 **ports_declaration** token in XML.

Channels

The metadata captured for a channel pertains to port level specifics and function signatures of the implemented access methods. For each access method encountered, we generate a **function_declaration** token similar to Figure 5.13 and for each interface port, we generate **port_declaration** as shown in Figure 5.14.

Polymorphic Components and Channels

We extract the template expression that describes a polymorphic entity and express its arguments as **template_parameter** tokens. These tokens capture three essential details about any argument besides the datatype and name. The attributes of a **template_parameter** token that embeds these details are:

1. *partial_spec* - Determines whether the argument is partially specified.
2. *constrained* - Determines whether the partially specified attribute is type constrained.
3. *position* - Captures the argument position in the template expression, which acts as the key for the processing and constraining performed on the component in stage 3.

In Figure 5.15, we show the **template_parameter** tokens used to capture the essentials of each argument in the template expression on the FIFO component in Figure 5.8.

```
<component name="fifo">
<template_parameter constrained="false"
datatype="class" name="DataT"
partial_spec="true" position="1">
<template_parameter constrained="false"
datatype="unsigned int" name="size"
partial_spec="false" position="2"/>
</component>
```

Figure 5.15 **Template_parameter** token.

The following observations about the FIFO can be made from this example: (1) FIFO is polymorphic with respect to the data it stores (DataT and partial_spec="true") and (2) it is a generic implementation with no specified constraints (constrained="false").

During extraction of polymorphic ports, the **port_declaration** token captures the template argument and its datatype and uses the partial_spec attribute to state that the port transmits/receives a template type. Extracting interface functions requires worrying about polymorphic function arguments or return-types (Figure 5.10) and is dealt in a similar manner as polymorphic ports. If the submodules of a module are templated, then during the extraction of the module hierarchy, the corresponding **template_parameter** tokens are generated.

General Extraction

There are high-level constructs that need to be mined to identify and extract the hidden metadata. Figure 5.16 illustrates the usage of *typedef* and its token representation in XML, where the attribute access_modifier captures the type of construct.

Comments

The corresponding XML token generated for a type-constraint is called the **constraint_entry** token. It consists of a set of **argument** tokens and identifies with the

```
// High-level construct
typedef sc_int<16> DataT;
// Type_declaration token
<type_declaration access_modifier="typedef"
bitwidth="16" data_type="sc_int" name="DataT"
template_type="false"/>
```

Figure 5.16 **Type_declaration** token for *typedef*.

template arguments through an attribute that captures its position in the template expression. For each **argument**, we insert a **dt_entry** token that captures the data type and bitwidth for all the legal types allowed through the constraint. Note that in Figure 5.17, the second argument of the template expression is fully specified and is mapped to "NA" in the constraint implying that the constraint has no bearing on the second argument.

```
//constraint : fifo <int; sc_int<32>; sc_int<64>, NA>

<constraint_entry name="C1">
    <argument position="1">
        <dt_entry type="int"/>
        <dt_entry length="32" type="sc_int"/>
        <dt_entry length="64" type="sc_int"/>
    </argument>
</constraint_entry>

//An entry in the constraint database
<component name = "fifo">
    <c_entry id = "C1"/>
</component>
```

Figure 5.17 Illustration of comment-level constraint and corresponding **constraint_entry** token.

5.3.3 Stage 3: Processing and Constraining DOM

This is a two phase stage with the populated **sc_DOM** given as input. The **sc_DOM** can be abstractly seen as a collection of components, channels, and interfaces. In this stage as shown in Figure 5.18, some of the **sc_DOM** constituents undergo a processing that results in updating/modifying some of the tokens. The processing is followed by a constraining phase where the stored implementations are type-restricted by propagating the effect of the specified constraint on the generic aspects of the implementation. The output of this stage is an IP library (DOM) called **cDOM**.

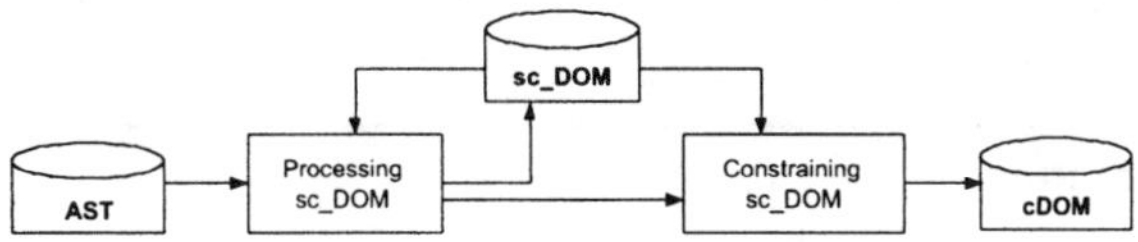

Figure 5.18 Stage 3: Processing and constraining sc_DOM.

Phase 1 - Processing sc_DOM

In this phase, we identify some ignored information from the AST parsing stage that requires analyzing the contents of the sc_DOM. Second, we process the sc_DOM to remove any indirection that hides the structural aspects of the design through placeholders. The discarded information that try to revive in this phase pertain to next-level interfaces. As mentioned in stage 2, we only identify first-level interfaces; however it is common to have next-level interfaces. Figure 5.19 shows a second-level interface *rdwr_if* that inherits first-level interfaces *read_if* and *write_if* of Figure 5.4.

```
//Inherits 1st level interfaces read_if and write_if
class rdwr_if : public read_if, public write_if { ... }
```

Figure 5.19 Second-level interfaces.

In stage 3, when we encounter a class element during the parsing of the AST, we try to identify it as a second-level interface. If successful, then the corresponding **interface** token is generated and appended to the sc_DOM. It is necessary to identify both first- and second-level interfaces, for the correct extraction of the TL behavior of a channel, the reason being that the key step in the extraction of a channel is to separate interface access methods from internal function definitions, which would require knowledge of all the interfaces that this channel implements. Therefore, after the extraction of the second-level interfaces, we perform another round of extraction for the channels that implement this interface. This results in the addition of more **interface_function** tokens to the **channel** token.

An interesting point to note is that second-level interfaces can give rise to third-level interfaces and so on. This unidirectional chain can continue to any depth and therefore we develop the extraction procedure shown below. It helps in navigating the chain and extracting the interfaces correctly. SystemC only allows for interfaces through safe usage of multiple inheritance; therefore, no cyclic dependency exists in the chain.

Interface Extraction

{Given **C** set of all classes and **IF** = **OC** = ϕ}
Step 1 For each $c \in$ **C**,
  Step 1.1 If c inherits *sc_interface* then insert(**IF**,c) else insert(**OC**,c)
Step 2 For each $oc \in$ **OC**
  Step 2.1 If oc inherits an interface $intf$ s.t. $intf \in$ **IF** $\wedge$ $oc \notin$ **IF**,
    then insert(**IF**,oc)

```
//Type indirection
typedef sc_int<16> DataT;

// IO Port
sc_in<DataT> in_real;
```

```
// Type_declaration token
<type_declaration
access_modifier="typedef"
bitwidth="16"
data_type="sc_int" name="DataT"
template_type="false"/>

// Port_declaration token
<ports_declaration
IOtype="sc_in" bitwidth="16"
datatype="DataT name="in_real"
partial_spec="false"/>
```

Figure 5.20 Example of type indirection and tokens generated.

Consider the following code snippet in Figure 5.20; the tokens shown are generated at the end of stage 2. The datatype extracted for the port is *DataT*, which basically is an alias for the actual type. Therefore to update such incorrect extractions, we apply the type-replacement procedure on the sc_DOM that removes all type indirections. Some of the structural entities that are affected by type indirection are:

- IO and interface ports;
- Interface function arguments and return-types;
- Template expression and template arguments;
- Submodule instantiations.

The replacement procedure runs through these entities searching for all indirection and replacing them with the actual types. This problem is further convoluted due to the same deep chain problem associated with interfaces. The type indirection can be repeated to any level, which requires iterating through the complete chain to find the actual type. The replacement procedure takes this problem into account and through an iterative scheme manages to replace all the type indirections at any depth with their respective datatypes. The procedure is shown next.

Replacement Procedure

{Given
T an ordered set of all type indirections,
P set of all IO/interface ports,
I set of all interface functions,
E set of all template expressions,
S set of all submodule instantiations}
Step 1 For each type indirection $t_i \in \mathbf{T}$,
Step 2 For each type indirection $t_j \in \mathbf{T}$,
  Step 2.1 If **search**(t_j, t_i) = *true* then **replace**(t_j, t_i)
Step 3 For each port $p \in \mathbf{P}$,
  Step 3.1 If **search**(p, t_i) = *true* then **replace**(p, t_i)
Step 4 For each interface $n \in \mathbf{I}$,
  Step 4.1 If **search**(n, t_i) = *true* then **replace**(n, t_i)
Step 5 For each template expression $e \in \mathbf{E}$,
  Step 5.1 If **search**(e, t_i) = *true* then **replace**(e, t_i)
Step 6 For each submodule instant $s \in \mathbf{S}$,
  Step 6.1 If **search**(s, t_i) = *true* then **replace**(s, t_i)

This simple procedure works well because of the order in which the **type_declaration** tokens are generated. In C++ to insert a typedef instance, the statement has to be valid; therefore the order in multilevel typedef-ing is very important and necessary for this procedure to terminate. We generate tokens and populate

T in the same order in which the typedefs are specified in the implementation. In Figure 5.21, we make use of a 2-D templated FIFO example to illustrate the application of the replacement procedure. The outcome of this phase is the updated **sc_DOM**, which is more complete and correct in terms of metadata.

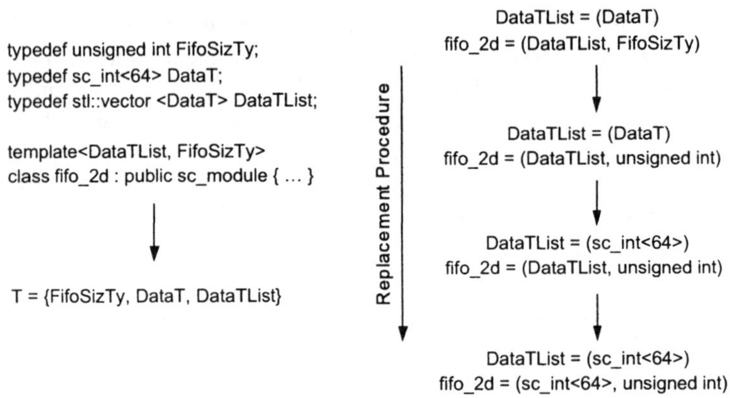

Figure 5.21 Application of the replacement procedure.

Phase 2 - Constraining sc_DOM

We discussed the extraction of comment-level constraints that a library engineer specifies to restrict the genericness of the IP. In this phase, these constraints captured as *constraint_entry* tokens are propagated through the IP. The propagation confines some polymorphic structural aspects of the IP to a set of legal types specified in the constraint.

In Figure 5.17, we had illustrated a comment-level constraint on the storage type of the FIFO in Figure 5.8. The result after propagating the constraint is shown in Figure 5.22, which appends to the **template_parameter** token for DataT, a set of **dt_entry** tokens that capture the possible types for substitution.

The polymorphic aspects of an IP range from IO ports to structural hierarchy. Therefore, the constraint propagation would require appending all of these aspects with the appropriate **dt_entry** tokens. The outcome of this phase is the **cDOM**.

```
<template_parameter constrained="true" datatype="class"
name="DataT" partial_spec="true" position="1">
    <dt_entry type="int"/>
    <dt_entry length="32" type="sc_int"/>
    <dt_entry length="64" type="sc_int"/>
</template_parameter>
```

Figure 5.22 Constrained **template_parameter** token.

5.4 IP SELECTION

We define the selection problem as follows:
"Given a CCM and cDOM of SystemC IPs, the selection problem is to identify at least one concrete implementation from the cDOM for every abstract entity in the CCM and create an executable model."

The primary task of the MCF is to select an appropriate IP from the library that can be plugged into the architectural template of the system to create possible executable models that allow for rapid exploration of design alternatives. We propose techniques that perform automated abstraction and design visualization-based matching between the virtual components in the architectural template of the system and the implementations in the IP library.

There are many ways to search for a suitable implementation, but the degree of suitability is often questioned. The best case would be if the virtual component has a one-to-one correspondence to an implementation that could be identified, but this is a difficult problem. It requires a more complete specification, which is tedious and defeats the purpose of a CCF. The average case is a partial identification, where the match is performed based on some common characteristics of the two. These characteristics are partly for convenience and partly for retrieval. We outline some of the characteristics that form the basis of our selection techniques. We also have mixed modes, where a selection is performed based on more than one characteristic. This is computationally more intensive, but reduces the search space and helps in attaining a closure on the set of matches obtained. We refer to the CCM entity being searched as the *door* and every implementation that matches as its *keys*. These terminologies allow the reader to quickly relate to whether we are referring to the abstract entity being searched or its corresponding matching implementation. The toplevel outline of the selection procedure is presented next, which is guided by the different proposed selection criteria.

Selection Procedure

{Given **C** a CCM model, **L** a cDOM of SystemC IPs and $\mathbf{M} = \phi$}
Step 1 For each component/media $c \in \mathbf{C}$,
  Step 1.1 For each implementation $l \in \mathbf{L}$,
    Step 1.1.1 If **match**(c, l, "selection_criterion") then insert(**M**, l)

Criterion 1 - Naming-Based

This criterion allows a quick selection of implementations that match based on their nomenclature. However, the degree of suitability is very low, since there could be many distinct implementations that follow similar naming conventions. The naming-based selection criterion is for convenience and is commonly used by modelers who are better informed about the IP library. For example, a modeler is looking for a 2-input, 2-output adder and is aware of the naming styles followed by the library engineer. The set of available adder implementations are adder_2_1, adder_2_2, adder_3_1, and adder_5_1. The modeler would have to name the virtual component in a similar fashion, so that the selection technique identifies the implementation of interest.

Criterion 2a - RTL Structure-Based

If the CCM and the library implementations are described through their IO behavior that structurally translates to input-output ports and signal-level connectivity, then we can perform RTL structure-based selection. In this case, the IP selector tries to match each port of the implementation to each port of the CCM. This functionality is performed by **type-match**, which returns a true, if the two are identical with respect to their datatype and bitwidth. If all the ports of the implementation match with all the ports of the CCM, then we have found one implementation that is structurally equivalent to the CCM and the **match** procedure returns a true.

---

match(c, l, RTL-based)

{Given **matchflag** = *false*}
Step 1 If **is-at-RTL**(l) and *c.portcnt* != *l.portcnt* then return *false*;
Step 2 For each port p_c of entity c,
  Step 2.1 There exist p_l port in implementation l such that,
    Step 2.1.1 type-match(p_c, p_l) = *true* else **'l' does not qualify**, return *false*;
Step 3 return *true*;

Criterion 2b - TL Structure-Based

If the CCM is described through read and write ports that structurally translates to interface ports and function calls, then we can perform TL structure-based selection.

An *interface* in SystemC is an abstract class and a *channel* inherits one or more of these abstract interfaces and provides implementation for all access function members of the interface. A *TL component* communicates via channels bound through interface ports. The IP selector has to match every abstract entity in the CCM with either a TL component or a channel. Therefore, the task of the selector can be outlined as follows: (1) For TL component, it matches with respect to their interface ports, and (2) for a channel, it matches with the interface functions that the channel implements. However, a channel can communicate with other channels similar to a component, through interface ports. Therefore, the IP selector has to also match with respect to interface ports to arrive at the complete match for a channel.

The outline of the selection algorithm is shown below. The program **if2if** is given two interface ports to see if they are structurally equivalent. Since the transaction behavior is a set of function calls, this equivalence checking boils down to comparing the function signatures of the access methods available through the two interface ports. If the program succeeds, then the **match** procedure returns a true. However for a channel, the selection performed by the **match** procedure is shown through step 2. The channel gives concrete definitions to the interface functions and two channels are equivalent if they define structurally similar access methods. However, a channel could hook up to a second channel through interface ports that would give it visibility to the functions in the second channel's structural definition.

The **match** procedure performs the functionality given by **if2if** for the channel-to-channel communication scenario and **fn2fn** for the normal scenario and tries to establish structural equivalence.

---

match(c, l, TL-based)

{Given **matchflag** = *false*}
Matching a component
Step 1 If **is-at-TL**(l) and !**is-channel**(l) then,
  Step 1.1 For each interface port p_c of entity c,
    Step 1.1.1 There exist an interface port p_l in implementation l such that,
      Step 1.1.1.1 if2if(p_c, p_l) = *true* else **'l' does not qualify**, return *false*;
  Step 1.2 return *true*;
Matching a channel
Step 2 else if **is-channel**(l) then,
  Normal Scenario
  Step 2.1 For each interface function f_c of entity c,
    Step 2.1.1 There exist interface function f_l in implementation l such that,
      Step 2.1.1.1 fn2fn(f_c, f_l) = *true* else **'l' does not qualify**, return *false*;
  Channel-to-Channel Scenario
  Step 2.2 For each interface port p_c of entity c,
    Step 2.2.1 There exists an interface port p_l in implementation l such that,
      Step 2.2.1.1 if2if(p_c, p_l) = *true* else **'l' does not qualify**, return *false*;
  Step 2.3 return *true*;

Mixed Mode Selection

The structure-based criteria ranks higher in terms of suitability. However, it is not the best possible criterion, as multiple unrelated implementations can have identical structural characteristics. A point to note is that our structure-based criterion does not consider hierarchy metadata while performing IP matching. Hierarchy information enforces strong requirement on the structure, which is not necessary to find matching implementations. For example, consider an virtual component for 4-bit addition. Some of the common possible implementations make use of 2-bit or 1-bit hierarchical structures as shown in Figure 5.23. All these implementations qualify as a match for the virtual component and therefore the usage of hierarchy metadata for selection should be provided as a choice to the user.

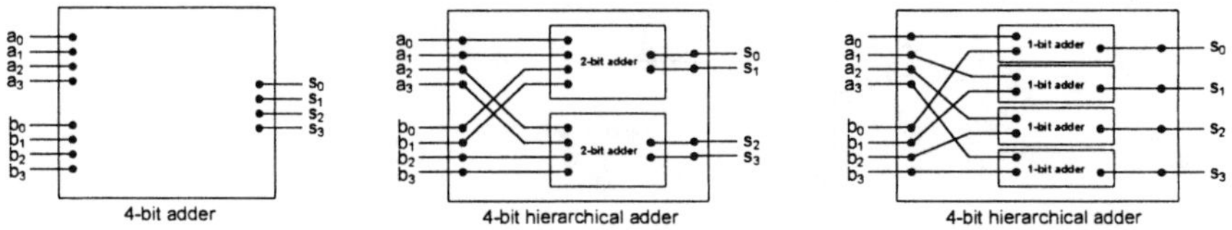

Figure 5.23 Various implementations of a 4-bit adder.

The criteria introduced so far assume the implementation to be a black-box with certain characteristics and ignore what the internals are and how they are arranged in the implementation. The black-box view of the *fir_top* component during selection is shown in Figure 5.24. We know from the previous subsection that we have the capability to reflect the structural hierarchy of an IP. This gives us visibility into how the different internals are patched together in the implementation. This information can be used to define points during the selection process. These points serve as demarcations for different visibility levels that guide the selection process into making more suitable choices for obtaining a match. The three visibility levels that we introduce are black-box view, flattened view, and hierarchical view. These views are mixed with our selection criterion, such that we use the visibility information with a particular distinguishing characteristic to better understand the IP being examined. This introduces the notion of mixed mode selection where every selection criterion (naming/structure-based) is provided with a possible view of the IP (black-box/flattened/hierarchical view). The reason to provide mixed mode selection is to allow various kinds of search, namely: (1) convenient search, (2) accurate search, and (3) search for any IP that is a probable match. The convenient search is for any expert user,[1] who has knowledge about the IPs in the IP library and wants to quickly pick up the IPs and put them together (naming-based + black-box view). The accurate search is for a user who wants to find the best IP that structurally matches the template (structure-based + hierarchical view). The search for any IP that is a probable match is meant for a novice user who does not have much knowledge about the IP library and is trying to explore the library to solve the design problem at hand (naming/structure-based + black-box/flattened).

The hierarchical view in a mixed mode selection is definitely the most suitable criterion. The black-box selection is the least suitable criterion and it is a superset of the selection obtained based on hierarchical viewing. The downside of performing

1. Implemented all the IPs in the library.

a hierarchical view based selection is the computational overhead, since a lot more aspects need to be equivalent to qualify as a match. The hierarchical view of the *fir_top* component is shown in Figure 5.25.

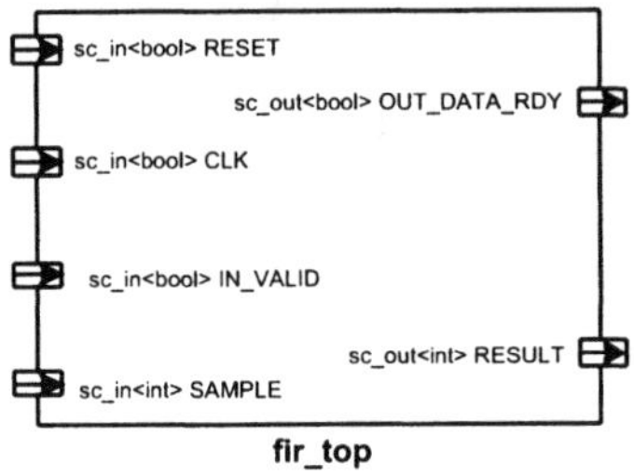

Figure 5.24 Black-box view of **fir_top**.

The flattened view is an alternate selection, where the implementation is partly treated as a black-box and partly by adding its internals into the library for consideration. For a hierarchical implementation, this selection criterion will ignore its internals (black-box treatment) while trying to match it to a virtual component. However, this selection criterion will increase the number of implementations considered for matching. This is because during such a selection the library is expanded to include the child implementations within an implementation, which basically means that the library will also contain the implementations that result from flattening all hierarchical implementations. The flattened view of the *fir_top* component is shown in Figure 5.26.

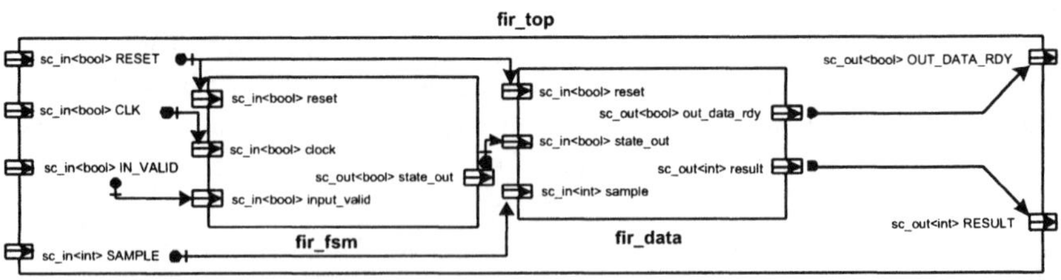

Figure 5.25 Hierarchical view of **fir_top**.

Our IP library allows for polymorphic IPs, which are generic with respect to the storage/transfer type. We also allow the implementer to provide a constraint file, in which they describe type-constraints on polymorphic parts of the IP. These are integrated into the library and consulted with during the selection. If no constraint file is provided, then the polymorphic component is generic enough to match with any virtual component at the datatype level. Note that the flattening technique requires IPs with polymorphic components to propagate the type constraints onto the polymorphic aspects. This enables an easy flattening of the complete IP without having to keep track of the possible dependencies in terms of type variables.

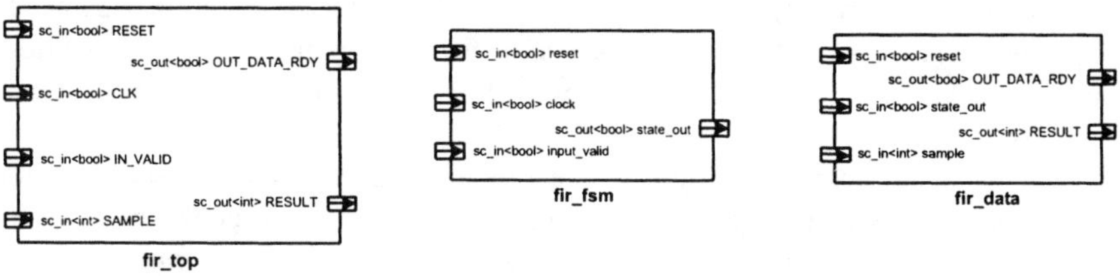

Figure 5.26 Flattened view of **fir_top**.

The main objective of IP selection is to arrive at a criterion that minimizes the match results, which necessitates a stronger matching characteristic. Future work would concentrate on behavioral types, where the implementation would be captured using a type-system and behavioral equivalence checking would be performed for selection.

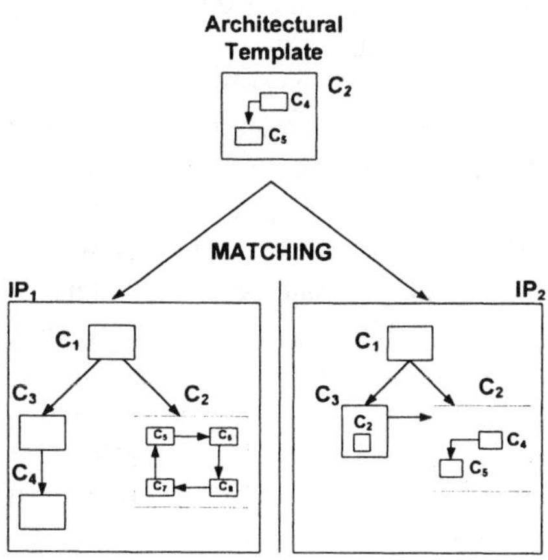

Figure 5.27 Example of an abstract component and two IP implementations.

The different selection schemes developed are summarized as: (1) for quick selections, we search based on nomenclature (version number, IP library name, and distributor), (2) for RTL components, we search based on parameter and port level structural type (data types), (3) for TL components and channels, we search based on interface-level structural type (function signatures and interface port specifics), (4) searching based on visualization of the IP, namely black-box, flattened, and hierarchical, (5) mix (1), (2), and (3) with opacity in (4) to obtain nine different selection schemes, and (6) an exhaustive search using the nine selection schemes to obtain the ideal match. The selection schemes take advantage of the composition-related metadata provided by the reflection and introspection (R-I) capability of the IP library.

In order to generate executables for the architecture template, the selection process should have succeeded. The match results need to be configured, where one of the many matching IPs will be picked based on certain power, area, and performance numbers and configured. Finally, the configured IPs are substituted into the template for generation of executables. Generation of verification models is the final task of MCF for the purpose of validating the IP composition. The verification models are SCV-based constrained/unconstrained random testbenches, which are also generated by introspecting the reflected metadata.

5.4.1 Illustrative Example

We illustrate naming-based IP selection on the example shown in Figure 5.27 by using the three views. In the example, an architectural template of $\mathbb{C}_2$ embedded with two subcomponents $\mathbb{C}_4$ and $\mathbb{C}_5$ is provided. The objective is to search the IP library consisting of two IP implementations (IP_1 and IP_2) for a match. We combine the naming-based criterion with the different views to enable the selection and the match results (MATCH_RESULT) obtained are shown in Figure 5.28.

The naming-based selection with black-box view (Figure 5.28(a)) searches for every implementation that blindly matches with the template $\mathbb{C}_2$. The implementation $IP_1.C_2$ obtained from IP_1 has a hierarchical structure that differs from the template $\mathbb{C}_2$, but the black-box view allows the selection to disregard the internals and consider $IP_1.C_2$ as a match for $\mathbb{C}_2$. The naming-based selection with flattened view (Figure 5.28(b)) obtains an additional match besides the ones obtained by the black-box selection. This match is the result of flattening component C_3 in IP_2, which has an internal component with similar naming conventions ($IP_2.C_3.C_2$). The final selection performed with hierarchical view (Figure 5.28(c)) results in eliminating one of the matches of the black-box selection. This match ($IP_1.C_2$)

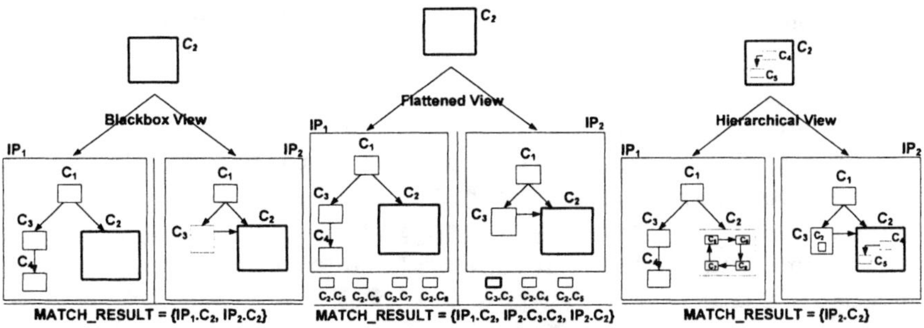

(a) Black-box selection (b) Flattened selection (c) Hierarchical selection

Figure 5.28 (a, b, c) Illustrating the naming-based selection with different views.

from IP_1 does not have a hierarchical structure that is equivalent to the structure of the architectural template C_2 and the selection eliminates it from the result set. Therefore, $(IP_2.C_2)$ is the ideal match for the template $\mathbb{C}_2$, which is obtained from IP_2.

```
<component file="fir_fsm.h" name="fir_fsm">
  <port_declaration IOtype="sc_in" bitwidth="1"
  datatype="bool" name="clock"/>
  <port_declaration IOtype="sc_in" bitwidth="1"
  datatype="bool" name="reset"/>
  <port_declaration IOtype="sc_in" bitwidth="1"
  datatype="bool" name="input_valid"/>
  <port_declaration IOtype="sc_out"
  datatype="unsigned int" name="state_out"/>
</component>
```

Figure 5.29 Metadata on the **fir_fsm** component.

5.5 CASE STUDY

Consider the FIR filter core from the SystemC distribution [1] at the RTL abstraction. It consists of a toplevel component, namely **fir_top** as shown in Figure 5.25, which hierarchical embeds the **fir_data** and **fir_fsm** components.

This **fir_top** is subjected to reflection and the metadata extracted is shown in Figure 5.29.

```
<component file="fir_data.cpp" name="fir_data">
  <port_declaration IOtype="sc_in" bitwidth="1"
  datatype="bool" name="reset"/>
  <port_declaration IOtype="sc_in"
  datatype="unsigned int" name="state_out"/>
  <port_declaration IOtype="sc_in"
  datatype="int" name="sample"/>
  <port_declaration IOtype="sc_out" datatype="int"
  name="result"/>
  <port_declaration IOtype="sc_out" bitwidth="1"
  datatype="bool" name="output_data_rdy"/>
</component>
```

Figure 5.30 Metadata on the **fir_data** component.

The **fir_fsm** component consists of three input ports and one output port as shown in Figure 5.29. The input ports are *clock*, *reset*, and *in_valid* of type Boolean. The output port is *state_out* of type unsigned integer. The **fir_data** component consists of three input ports and two output ports as shown in Figure 5.30. It also has a *reset* input of type Boolean and two other inputs, namely *sample* and *state_out* of type unsigned integer and integer, respectively. The toplevel component **fir_top** has four inputs and two outputs. It also encapsulates the two components **fir_fsm** and **fir_data** as subcomponents.

Next, let us consider the simple bus model from the SystemC distribution [1] at the TL abstraction. Figure 2.3 describes the TL model, which consists of three bus masters with distinct protocol-level interfaces (blocking, nonblocking, and direct read-write access) and an arbiter to schedule their transactions. It consists of two slaves, one with instantaneous read-write access and the other with slow read-write access due to its clock-based activation. The master components describe their interface read-write transaction through function calls, which are implemented through a bus channel. An arbiter channel implements the arbitration interface. The two slaves are also modeled as channels.

In Figure 5.31, we illustrate the metadata extracted for the nonblocking interface **simple_bus_non_blocking_if**. The interface consists of three function declarations, namely *read*, *write*, and *get_status*. The *read* function allows the

```
<interface name="simple_bus_non_blocking_if">
  <function_declaration name="read" return_type="void">
    <argument datatype="unsigned int"
    name="unique_priority"/>
    <argument datatype="int" name="data" ptr="true"/>
    <argument datatype="unsigned int" name="address"/>
    <argument bitwidth="1" datatype="bool" name="lock"/>
  </function_declaration>
  <function_declaration name="write" return_type="void">
    <argument datatype="unsigned int"
    name="unique_priority"/>
    <argument datatype="int" name="data" ptr="true"/>
    <argument datatype="unsigned int" name="address"/>
    <argument bitwidth="1" datatype="bool" name="lock"/>
  </function_declaration>
  <function_declaration name="get_status"
  return_type="simple_bus_status">
    <argument datatype="unsigned int"
    name="unique_priority"/>
  </function_declaration>
</interface>
```

Figure 5.31 Metadata on the nonblocking bus interface.

exchange of the following arguments: (1) *unique_priority*, (2) *data*, (3) *address*, and (4) *lock*, between the master and the bus. These arguments allow the bus to pass on the request to the corresponding slave. The *unique_priority* argument of type unsigned integer indicates the priority of the current request. The *data* is a pointer of type integer that contains the information read from the slave. The *address* field is used to help the bus figure out which slave is the target for the current request. Finally, the lock argument is used to indicate that the request cannot be preempted unless it is completed. Similarly, the function declarations of *write* and *get_status* are also shown in Figure 5.31.

The nonblocking master uses the nonblocking interface (Figure 5.31) to make requests on the bus, which is allowed through the port (*bus_port*) that binds it to **simple_bus_non_blocking_if**. The metadata extracted for the nonblocking master is

```
<component abstraction="TL"
file="simple_bus_master_non_blocking.h"
name="simple_bus_master_non_blocking">
<parameter datatype="sc_module_name"/>
<parameter datatype="unsigned int"
           name="unique_priority"/>
<parameter datatype="unsigned int"
           name="start_address"/>
<parameter bitwidth="1" datatype="bool" name="lock"/>
<parameter datatype="int" name="timeout"/>
  <port_declaration IOtype="sc_in" bitwidth="1"
  datatype="bool" name="clock"/>
  <port_declaration IOtype="sc_port"
  datatype="simple_bus_non_blocking_if"
  name="bus_port"/>
</component>
```

Figure 5.32 Metadata on the nonblocking master.

shown in Figure 5.32. The **parameter** tokens show the externals that can be set by a toplevel for the **simple_bus_master_non_blocking** component.

The blocking and direct interfaces and the blocking and direct masters are also extracted in a similar manner to the nonblocking interface and master. The metadata extracted on the bus channel that implements the nonblocking, blocking, and direct interfaces is shown in Figure 5.33. It has an input clock port as well as two interface ports, which allow it to interface with the slaves and the arbiter. The channels for the arbiter and the slaves are also extracted in a similar manner. Note that the **partial_spec** attribute is omitted from all the token illustrations in above case studies as it is always set to *false* for a nonpolymorphic component or channel.

5.6 SUMMARY

We have illustrated the extraction of introspective metadata from SystemC IPs. Our methodology combines the KarSCPar SystemC Parser, XML tools, and the DOM data structure to enable reflection and introspection of SystemC designs. The metadata serves as input to a CCF, which allows for IP selection and composition to create an executable given an abstract specification and a library of SystemC IPs.

Furthermore, we have outlined some of the challenges in terms of multistage processing required to mine metadata for composability of IPs from designs specified in C++.

```
<channel abstraction="TL" file="simple_bus.h"
         name="simple_bus">
<parameter datatype="sc_module_name"/>
<parameter bitwidth="1" datatype="bool" name="verbose"/>
  <port_declaration IOtype="sc_in" bitwidth="1"
  datatype="bool" name="clock"/>
  <port_declaration IOtype="sc_port"
  datatype="simple_bus_arbiter_if" name="arbiter_port"/>
  <port_declaration IOtype="sc_port"
  datatype="simple_bus_slave_if" name="slave_port"/>
  //interface implementation of blocking protocol
  ...
  //interface implementation of nonblocking protocol
  ...
</channel>
```

Figure 5.33 Metadata on the bus.

Given an abstract specification and a library of IPs, performing the different selection schemes proposed in this chapter allows for rapid exploration of design alternatives. It reduces the time to develop system models, which dramatically improves design quality. In this chapter, we have identified three basic criteria and three visibility modes that are mixed and matched to obtain nine different selection schemes.

References

[1] OSCI Group, SystemC Web site, http://www.systemc.org/.

[2] The Apache Software Foundation, "Xerces C++ validating XML Parser," http://xml.apache.org/xerces-c/.

Chapter 6

Typing Problems in IP Composition

The IP composition problem that we address exists in a design flow that ties in a metamodeling framework [1] for system specification at the architectural level and the necessary tool-support for automated reflection and introspection [2], selection [3], configuration, and composition. Such a design flow is used by MCF [4], a metamodeling-driven component composition framework for SystemC based IP composition. One of the main ingredients of MCF is its composition language that allows a modeler to construct an architectural template for system (IP) description. The *architectural specification* enabled through MCF promotes design exploration, reuse, and reduces specification time.

In high-level languages such as C++, IP components are described in a flexible manner so that they can be used in a wide range of designs. Such a description is practical [5], if the usage of flexible components is not associated with a high overhead. Common techniques used to describe flexible components are *component overloading* [6] or *parametric polymorphism* [7]. Component overloading alleviates the type compatibility problem in strongly typed languages such as C++ and promotes reuse by rendering the component somewhat independent of the context of usage. It provides flexibility, but requires several implementations; one for each data type. Another approach to describing flexible components is to make them independent of the data types they manipulate. Examples of such components would be memory, FIFOs, and switches, whose high-level functionality is not bound to any data type. C++ allows for such generic descriptions through parametric polymorphism in which a component is built using a template and uses type-variables that are placeholders for concrete types and resolved during instantiation. However, the excessive type instantiations needed discourage their usage.

During the architectural specification in MCF, a component's interface in terms of input-output ports or read-write calls is bound to a set of data types. These data types are collected to create a contract for the component. The contract describes type-bindings that are data types that the component's interface must conform to during its interaction. MCF also allows a *partial architectural specification* to enable the description of flexible components. During a partial specification, the component is made typeless by not binding it to any type-contract. However, if a component is bound to a type-contract, then its interactions are constrained by the type-bindings specified in the contract, thereby restricting its flexibility. The tool-support in MCF employs the architectural specification of the system to perform automated selection of matching implementations from an IP library. These are plugged into the architectural template for composition. If the components in the architectural specification are bound to type-contracts, then the composition problem simplifies into a selection problem resolved by data type-matching. However with flexible components, the automatic composition problem is nontrivial and requires type inference. Given an IP library, our type resolution algorithms infer the most specific solution that solves the type inference problem.

In this chapter, we outline the type inference problem in MCF and discuss our multistage solution that performs type-propagation and type-matching to convert a partial architectural specification into an architecture specification to facilitate IP composition. To informally state the problem, given an architectural specification of the system with flexible components, we want to perform automated IP composition through sound type inference techniques and create an executable specification.

6.1 MCF TYPE DEFINITIONS

MCF alleviates the tediousness associated with knowing the specifics of the software language and allows a hardware designer to realize a system through visual architectural templates and tool-support. The template structurally supports the modeling abstractions such as RTL, transaction level (TL), mixed-abstractions (RTL-TL), and structural hierarchy. Furthermore, the framework enforces checks that look for inconsistencies, data type mismatches, incompatible interfaces, and illegal component or channel descriptions. The MCF tool-support enables automatic selection, configuration, and substitution of IP-cores from an IP library into the template to arrive at various executables. Test generators to verify the IP composition are also provided as a part of MCF.

The three requirements to solve the MCF type inference problem are the architectural specification, the IP library, and the MCF type-system. The architectural specification is enabled through the component composition language (CCL) in MCF. The IP library is a collection of compiled SystemC IP-cores from the SystemC distribution, their structural variants, and industry contribution. The MCF type-system captures the relevant types of the target design language and their relationships. The type-system is given as input to the inference techniques for the purpose of type-matching. These techniques consult the type-system during type-checking to determine whether there is a type mismatch in the specification. The MCF type inference is a two-part problem, where the first part addresses the type inference issues at the architectural level with abstract components and the second part addresses the issues with replacing abstract components in the template with real implementations from an IP library. In this section, we formally define the terminologies necessary to comprehend the MCF type-inference problem.

6.1.1 Component Composition Language

The syntax and semantics for component composition in CCL are developed using a metamodeling framework [1], which brings forth the following advantages: (1) allows expressing constraints[1] that enforce modeling rules and (2) enables language extensions through easy customizability. The CCL consists of a set of entities that are provided as modeling constructs. The modeler is allowed to instantiate, characterize, and further connect these entities to describe the architectural template of the system. During characterization, the instantiated entities are annotated with composition-related metadata such as naming and type information. However, certain rules expressed as OCL constraints are enforced by the metamodeling framework and checked at design-time (instantiation, characterization, and creating connections).

An *entity* in the CCL could be instantiated as a *component* or a *medium*. Components are of two types. A *leaf* component represents an entity without any internals or embeddings. A *hier* component is a hierarchical entity that embeds other leaf or hier components. A *medium* is a communication primitive used to describe various interaction styles such as bus/channel-based or network-on-chip (NoC) as well as connectors such as splitters and mergers for (de-)multiplexing functionality. The interface of these entity instances differ based on the modeling abstraction.

1. These are rules enforced on entities and relations and articulated based on the predicate expression language called OCL [8]. In MCF, they are used to express association constraints among the entities, type-matching rules, function signature matching rules, containment hierarchy rules, as well as the static properties of the different entities in the CCL.

An instance at RTL describes their interface behavior through *input-output* (IO) ports and signal-level connectivity. An instance at TL dictates the transaction at the interface through read-write function calls and channels. In Figure 6.1, we illustrate the architectural template of an RTL adder and a TL FIFO. The template ADDER_2_2 has two inputs, namely in_1 and in_2 and two outputs, namely sum and $carry$. The template FIFO has a read interface port $fifo\_read$ and a write interface port $fifo\_write$.

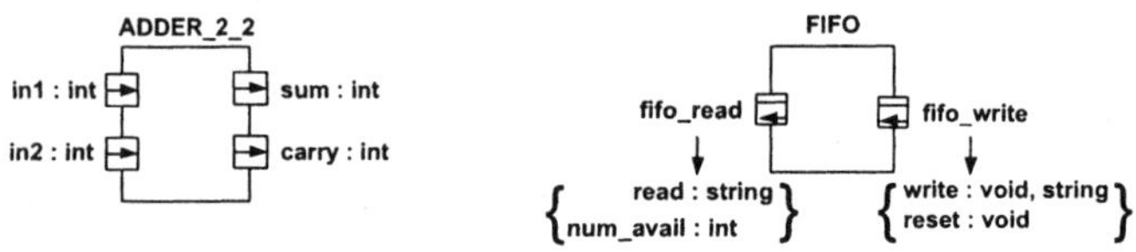

Figure 6.1 Illustration of architectural templates.

The *type-contract (TC)* is the collection of type-bindings strictly enforced on the various instances and their connection in the architectural template. For an RTL instance, the TC captures the port specifics, where each port is associated with a *data type* and *bitwidth* that describes the type of value exchanged and enforced through the binding as shown in Figure 6.1 for ADDER_2_2. An instance at TL is described using interface ports, which grants them access to interface-level read and write functions. Therefore, the TC consists of function signatures of all the access methods available via the interface port as shown in Figure 6.1 for the FIFO. A *function signature* captures the set of parameters associated with a method, where each parameter has a data type and bitwidth. The *read* method accessible via the *fifo_read* port exchanges one parameter which is of type `string`. The first parameter of any method is the return type, which is void for the *write* method accessible through the *fifo_write* port.

The partial specification capability of MCF allows for certain parts of the template to be unspecified and is denoted by $\perp$ in our formalism. The TC of an architectural template is formally defined here:

Definition 10. ***Type-contract***
The TC is represented as $\langle I, P, \mu_P, T_\perp, \gamma_P, \lambda \rangle$, *where*
I is a set of entity instances.
P is a set of ports associated with all instances.
$\mu_P : P \rightarrow I$ *is a function to map a port to the instance it belongs to.*
T is a set of all concrete data types (D) and all possible type-variables (V), that is, $T = D \cup V$.

γ_P: $P \rightarrow T_{\perp}^{n}$ is a function that maps a port to a type-vector (t_p), where each vector element is a data type, $T_{\perp} = T \cup \{\perp\}$ and $n \in N$.
$\lambda \subseteq P \times P$ is a set of valid connections.

Each port has a data type. The data type of the port restricts the type of the data values that can pass through it. These types are specified by the modeler, or left unspecified/partially specified for one of the following reasons: (1) rapid specification or (2) flexible components. The unspecified data types are denoted as $\perp$. The ports that are connected through a common connection must have identical data type-bindings. If these data types are unspecified, then we denote them by a type-variable. All the $\perp$s and the type-variables are solved during type resolution. The ports that have the same type-variable as their data type declaration must be resolved to have the same type.

An IO port $p \in P$ has a type t that is captured by a single element type-vector $t_p = \langle t \rangle = t$ (for simplicity) identified via a γ_P mapping (t $\in T_{\perp}$).

Definition 11. *An IO port is untyped if $p_{io} \in P : \gamma_P(p_{io}) = \perp$. It is fully typed if $p_{io} \in P : \gamma_P(p_{io}) = t$, where $t \in D$. It is type-constrained if $p_{io} \in P : \gamma_P(p_{io}) = t$, where $t \in V$.*

The interface ports in a TL instance serve as a collection of variable names, where they facilitate transfer of data values. Therefore, the type-vector t_p of an interface port $p \in P$ with n variables that contains a type for each value variable is $t_p = \langle t_1, t_2, \ldots, t_n \rangle$, where $t_i \in T_{\perp}$.

Definition 12. *An interface port is partially typed if $p_{if} \in P : \gamma_P(p_{if}) = \langle t_1, t_2, \ldots, t_n \rangle$, where $\exists t_i$, ($t_i = \perp$) $\vee$ ($t_i \in V$).*

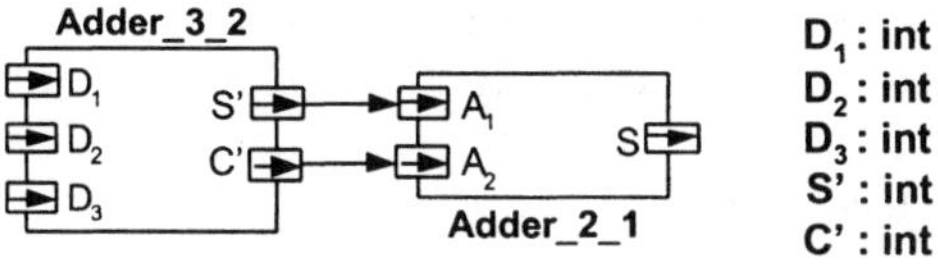

Figure 6.2 An abstract example.

Consider the architectural specification of an Adder_3_2 connected to an Adder_2_1 shown in Figure 6.2. The TC captures the following:

I = {Adder_3_2, Adder_2_1}
P = $\{D_1, D_2, D_3, S', C', A_1, A_2, S\}$

$\mu(D_1) = \mu(D_2) = \mu(D_3) = \mu(S') = \mu(C') = \text{Adder\_3\_1}$
$\mu(A_1) = \mu(A_2) = \mu(S) = \text{Adder\_2\_1}$
$\gamma_P(D_1) = \gamma_P(D_2) = \gamma_P(D_3) = \gamma_P(S') = \gamma_P(C') = \texttt{int}$
$\gamma_P(A_1) = \gamma_P(A_2) = \gamma_P(S) = \perp$
$\lambda = \{(S', A_1), (C', A_2)\}$

Each connection is used to link two ports of distinct/same instances. Every valid port-to-port connection $(p_i, p_j) \in \lambda$, satisfies a set of association constraints and static requirement based constraints. The **association constraints** are as follows:

1. $\forall\, p_i \in P, (p_i, p_i) \notin \lambda$

2. $\forall\, p_i, p_j \in P$, If $(p_i, p_j) \in \lambda$ then $(\gamma_P(p_i) = \gamma_P(p_j)) \vee (\gamma_P(p_i) = \perp) \vee (\gamma_P(p_j) = \perp)$

These association constraints enforce the connection validity, which is restricted to matching types or their types being unspecified ($\perp$).

The allowable characterizations and possible containments of an object and the relations that it can participate in defines its static requirement that is strictly followed by the modeler. Each entity in the metamodel has a specific set of properties that are mapped to **static-requirement constraints**, which are enforced by the metamodeling framework during modeling. In a very abstract sense, they can be thought of as constraints on association constraints. They constrain two different port-connections of an instance (that may or may not share ports) based on the properties of the instance. Constraints A and B are examples of static-requirement constraints.

For example, a bus instance is allowed to have an input port driven by multiple output ports for a request arbitration scenario as shown in Figure 6.3 on the left. A constraint enforced by the framework, while connecting the components $master_1$ and $master_2$ to the bus b is shown here:

- **Constraint A:** If $(p_1, p_b) \in \lambda$ for $\mu(p_b) = b$, then $(p_2, p_b) \in \lambda$ iff $(\gamma_P(p_1) = \gamma_P(p_2)) \vee (\gamma_P(p_1) = \perp) \vee (\gamma_P(p_2) = \perp)$

Constraint A enforces a rule on (p_1, p_b) and (p_2, p_b), which says that both components communicating through a common bus port must have their ports bound to the same type or unspecified. In this example, the ports p_1 and p_2 are not connected (i.e., $(p_1, p_2) \notin \lambda$, but the Constraint A applies an association-constraint between them).

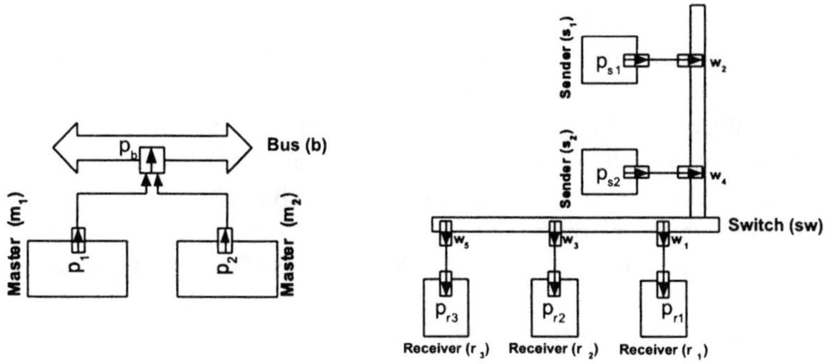

Figure 6.3 Examples illustrating properties of a bus (b) and a switch (sw).

Consider another example of two senders communicating with three receivers over an NoC using a common switch line as shown in Figure 6.3 on the right. The senders s_1 and s_2 attach themselves to the switch line (sw) through their respective ports p_{s_1} and p_{s_2}. Similarly, the receivers r_1, r_2, and r_3 attach themselves to sw through p_{r_1}, p_{r_2}, and p_{r_3}, respectively. The switch ports are w_1, w_2, w_3, w_4, and w_5, where w_2 and w_4 are used by the senders and the rest by the receivers. A constraint enforced by the framework, while connecting the senders (s_1 and s_2) and receivers (r_1, r_2, and r_3) to the switch line is shown here:

- **Constraint B:** If $(p_{s_1}, w_2) \in \lambda$ for $\mu(w_2) = sw$, then
  $(p_{s_2}, w_4) \in \lambda$ iff
  $(\gamma_P(p_{s_1}) = \gamma_P(p_{s_2}) = \gamma_P(w_2) = \gamma_P(w_4) \neq \bot)$ (1)
  $\vee ((\gamma_P(p_{s_1}) = \gamma_P(p_{s_2}) = \gamma_P(w_2) \neq \bot) \wedge (\gamma_P(w_4) = \bot))$ (2)
  $\vee ((\gamma_P(p_{s_1}) = \gamma_P(p_{s_2}) = \gamma_P(w_4) \neq \bot) \wedge (\gamma_P(w_2) = \bot))$ (3)
  $\vee ((\gamma_P(p_{s_1}) = \gamma_P(w_2) = \gamma_P(w_4) \neq \bot) \wedge (\gamma_P(p_{s_2}) = \bot))$ (4)
  $\vee ((\gamma_P(p_{s_2}) = \gamma_P(w_2) = \gamma_P(w_4) \neq \bot) \wedge (\gamma_P(p_{s_1}) = \bot))$ (5)
  $\vee ((\gamma_P(p_{s_1}) = \gamma_P(p_{s_2}) \neq \bot) \wedge (\gamma_P(w_2) = \gamma_P(w_4) = \bot))$ (6)
  $\vee ((\gamma_P(w_2) = \gamma_P(w_4) \neq \bot) \wedge (\gamma_P(p_{s_1}) = \gamma_P(p_{s_2}) = \bot))$ (7)
  $\vee ((\gamma_P(p_{s_1}) = \gamma_P(w_2) \neq \bot) \wedge (\gamma_P(p_{s_2}) = \gamma_P(w_4) = \bot))$ (8)
  $\vee ((\gamma_P(p_{s_1}) = \gamma_P(w_4) \neq \bot) \wedge (\gamma_P(p_{s_2}) = \gamma_P(w_2) = \bot))$ (9)
  $\vee ((\gamma_P(p_{s_2}) = \gamma_P(w_4) \neq \bot) \wedge (\gamma_P(p_{s_1}) = \gamma_P(w_2) = \bot))$ (10)
  $\vee ((\gamma_P(p_{s_2}) = \gamma_P(w_2) \neq \bot) \wedge (\gamma_P(p_{s_1}) = \gamma_P(w_4) = \bot))$ (11)
  $\vee (\gamma_P(p_{s_1}) = \gamma_P(w_4) = \gamma_P(p_{s_2}) = \gamma_P(w_2) = \bot)$ (12)

Constraint B is specified based on the assumption that sender s_1 was the first component connected to the switch. The constraints are generated in the order in which the other components are connected, namely s_2, r_1, r_2, and r_3. Constraint B shows the 12 rules, one of which must be true for sender s_2 to be connected to sw or else a violation is flagged. Similar rules are generated when the receivers are connected on to sw. These constraints state that all components communicating through a common switch must have their ports bound to the same type or unspecified. Note that sw is a *1-type* switch, where only a single type of data is communicated. However, the CCL allows the modeler to create different switch configurations such as the *n-m-type* switch, where "n" different types of data are communicated, such that each type is communicated between "m" components. Therefore, the switch allows (n × m) components to communicate through the same switch. A constraint on such a switch groups the components that must communicate using the same data type.

Definition 13. *An RTL instance* $i \in I$ *is partially typed if* $\exists p_{io} \in P,\ \mu_P(p_{io}) = i :$ $\gamma_P(p_{io}) = t$, *where* $(t = \perp) \vee (t \in V)$. *It is fully typed if* $\forall p_{io} \in P,\ \mu_P(p_{io}) = i :$ $\gamma_P(p_{io}) = t$, *where* $t \in D$.

6.1.2 IP Library

Given a library of SystemC designs to perform IP composition, we need techniques to perform reflection and introspection. *Introspection* is the ability of an executable system to query internal descriptions of itself through some reflective mechanism. The *reflection* mechanism exposes the structural and runtime characteristics of the system and stores it in a data structure. The data stored in this data structure is called *metadata*. The details on the metadata extraction by MCF from the SystemC IP library is provided in [2].

The metadata on an RTL implementation consists of information on input-output ports (sc_in/sc_out) and clock ports (sc_in_clk/sc_out_clk). For these ports, information related to the data type and bitwidth are essential constituents of the metadata. For an implementation at TL, the metadata contains the information of interface ports (sc_port), clock ports, and interface access methods. Therefore the metadata would contain the function signatures that embed the function arguments and return types. The hierarchical structure of a component is also essential for IP composition. In Figure 5.11, we illustrate the result obtained after performing reflection on the RTL FIR filter model from the SystemC distribution [9]. The port-level specifics and signal connectivity extracted as a part of the metadata is also shown. Another class of metadata that enables integration of semiconductor IP and IP tools is called *interoperability metadata* and SPIRIT [10] illustrates an SoC design flow by utilizing this metadata exchanged in a design-language neutral format.

Note that the SystemC IP library contains both polymorphic as well as overloaded IPs. For polymorphic components and channels in the library, MCF reflects the C++ template information and for overloaded IPs, the possible type instantiations that restricts the IP usage. We denote polymorphic aspects of an IP using $\sqcup$ and the *reflected structural type (RT)* of an IP is formalized here:

Definition 14. ***Reflected Structural Type***
The RT is represented as $\langle M, P, \mu_P, T_\sqcup, \gamma_P, R, \alpha_P \rangle$*, where*
M contains all the components and channels in the implementation of the IP.
P is a set of ports associated with all components and channels of the IP.
$\mu_P : P \rightarrow M$ *is a function that maps a port to the component or channel of the IP it belongs to.*
T is a set of all concrete C++ data types.
R is a set of user-specified IP type-restriction.
γ_P*:* $P \rightarrow T_\sqcup^n$ *is a function that maps a port to a type-vector, where each element is a C++ data type,* $T_\sqcup = T \cup \{\sqcup\}$ *and* $n \in N$*.*
α_P*:* $P \rightarrow R$ *is a function that maps a port to an IP type-restriction.*

M is referred to as an *implementation*[2] and can be used to substitute an entity instance in the architectural template. P is the set of IO or interface ports, which are similar to ports in the architectural template, with the difference that they can be polymorphic. R consists of type-restrictions that are used to constrain polymorphic ports of an implementation. A restriction $r \in R$ specified on a polymorphic port $p \in P$, describes the type instantiation choices that are allowed. It permits p to be instantiated by any data type from the choices presented by r. A restriction r has the following format $r := t_1|t_2|\ldots t_m$, where $t_i \in T$ and can be used for a type-assignment of p's data type and bitwidth.

Definition 15. *A reflected IO port of an IP is overloaded if* $p_{io} \in P : \gamma_P(p_{io}) = \sqcup$ *and* $\alpha_P(p_{io}) = r, r \in R$*. It is polymorphic if it is only mapped to* $r_\sqcup$ *(*$r_\sqcup = \sqcup$*).*

This intuitively means that the polymorphic port can be instantiated with any legal C++ data type.

6.2 TYPE RESOLUTION IN MCF

The type inference problem in MCF begins with a partial specification. Allowing for partially typed components and media in the architectural template brings forth the

2. An implementation consists of components or/and channels in a SystemC IP.

following advantages: (1) promotes reusability and a fast specification process, and (2) avoids the tediousness associated with characterizing every instance, especially when automatic inference can be performed. The types are inferred based on their static requirement and from the interactions with their neighboring instances, and also allows for polymorphic implementations in the IP library, which can also be selected besides the other library elements to substitute partially typed instances and derive an executable system model.

Definition 16. ***MCF Type Inference Problem***
Given an untyped or partially typed architectural template with a set of instances interconnected through port-connections, the problem is to derive a fully specified type-contract for the architectural template such that each instance conforms to their respective type-bindings.

This type inference problem is resolved using our two-stage algorithm. In the first stage, the type inference is performed on the architectural template, where the type-bindings are derived from association-based propagation and the static nature of the communicating instances. If the outcome of this stage is not a fully specified TC for the template, then we proceed to the next stage of the algorithm for further derivations. In the second stage of the algorithm, we make use of the IP library, where the type inference problem is converted into a type substitution problem. This problem requires selecting an implementation that satisfies the type-bindings specified by the TC of the architectural template. The RTs of the different implementations in the library are used to substitute the type-variables in the partially bound contract and complete the type inference problem.

6.2.1 Type Inference on Architectural Template

The type inference performed in the first stage is done by applying Algorithm 1.

Algorithm 1: Type Inference on Architectural Template

{Given an unspecified/partially specified type-contract TC = $\langle I, P, \mu_P, T, \gamma_P, \lambda\rangle$, a `violation` flag set to false.}
Step 1 For each $i \in I$,
//`assoc-based-derivation`
Step 1.1 For each $p_j \in P$, **such that** $\mu_P(p_j) = i$
If (*untyped*(p_j) $\vee$ *partially typed*(p_j))
then

For each $c \in \lambda$, **such that** $c = (p_j, p_k)$ where $p_k \in P$
If (!*untyped*(p_k) $\wedge$!*partially typed*(p_k))
then
type-assignment(p_j, p_k); break;
Else
type-variable-assignment(p_j, p_k); break;
If (*consistency-check-failed*(TC) = true)
then
`violation` = *true*; break;
End for_loop
End for_loop
Step 1.2 Perform *req-based-derivation(i)*;
If (*consistency-check-failed*(TC) = true)
then
`violation` = *true*;
End for_loop

Application of the algorithm to an architectural template results in one of the three outcomes:

- Case 1: A fully specified type-contract is obtained, which implies that the inference problem has been resolved.
- Case 2: An unspecified/partially specified type-contract is obtained, then its configuration serves as input for the second stage.
- Case 3: The third scenario is the undesirable case, when an error is encountered during type inference.

The following explanation is provided with respect to untyped ports and is similar for partially typed ports.

Given the instance set, the algorithm begins with an instance and tries to infer the type-bindings of its untyped ports from its connections. We call this approach the *association* based derivation (*assoc-based-derivation*). It is performed on untyped ports or partially typed ports, where the port has one or more elements in its type-vector that map to $\perp$. The result is a *type-assignment*, where the type-vector of an untyped port is updated with the type-vector of the connecting fully typed port. If the algorithm cannot infer the type and perform a type-assignment for an untyped port, then it performs the next step, a *type-variable-assignment*, during which both ports participating in the connection are mapped to a type-variable instead of $\perp$. A type-variable assigned to a port-connection enforces that both ports must have identical type-bindings.

Upon completion of *association* based derivation, we apply the *static-requirement* based derivation (*req-based-derivation*), where a certain property of an instance guides the inference process by relating two or more of its ports that are not connected to each other in the template. This requires the instance to have some initial type-bindings, which the algorithm uses to derive other type-bindings and performs assignments. The outcomes possible after the application of the req-based-derivation are: (1) some/all type-variables assigned to the ports of the instance (as a result of the association-based derivation) are replaced by an inferred type that was derived from one or more fully typed ports of the instance, or (2) some/all type-variables assigned to the ports of the instance are replaced by a new type-variable.

The algorithm terminates when it has completed one iteration over the set of instances in the template. For an instance $i \in I$, the algorithm performs association-based derivation followed by static-requirement based derivation. Upon completion of these two derivations, we say the type inference on i is complete and the status of the type-vectors of its ports in TC is called its *final configuration*. If the final configuration of i has all its ports fully typed, then the algorithm has managed to arrive at a fully specified derivation for the instance. If the final configuration of i has at least one port that is partially typed, then the algorithm has not managed to fully specify the instance and the outcome is a partially specified instance.

At the end of the iteration, if all instances in the template are fully typed, then the algorithm has obtained a fully specified type-contract (Case 1). If there exists an instance in the template that is partially specified at the end of the iteration, then the algorithm has obtained a partially specified type-contract (Case 2). If the TC obtained at the completion of the algorithm is partially specified, then it is used as input for the second stage of the type inference problem. Note that at this point every port in P is either typed or mapped to a type-variable.

If a type-assignment performed by the algorithm violates any **association** or **static-requirement constraint** enforced by the metamodeling framework, then the `violation` flag is set (Case 3). This happens during inference, when a type was propagated and assigned to an untyped/partially typed port causing an inconsistency with respect to the other fully typed ports of the instance. The function used by the framework to detect any type violations in terms of mismatches in the architectural template is the *consistency-check-failed*. This function employs the type-system to check whether the two data types of the ports being matched as a result of an enforced constraint are either *type equivalent* or *type subsumed*. If the two data types fail to conform to either, then the `violation` flag is set. At this point, the algorithm terminates by stating that the architectural template is not correct and there is a type inconsistency with respect to the interface specification of certain instances in the template. The designer needs to go back to the architectural template and rework the type-contract to remove the conflicting scenario.

The time-complexity of the algorithm is $O(|I| \times (|\lambda| + |\lambda|)) = O(|I| \times |\lambda|)$. The outermost loop of the algorithm runs for $|I|$ iterations. The *assoc-based-derivation* as well as the *req-based-derivation* are performed at most $|\lambda|$ times each before the algorithm terminates. The *assoc-based-derivation* analyzes the two ports participating in the connection and performs a *type-assignment* or a *type-variable-assignment*. The *req-based-derivation* analyzes the set of connections of a instance based on its properties and tries to unify the type-variables of the connections through a new *type-variable-assignment* or performs a *type-assignment* per connection.

6.2.2 Type Substitution Using IP Library

The output of stage 1 is an architectural template with a TC that consists of type-variables constraining port-connections. In stage 2, these type-variables are substituted by C++ data types using RTs of the implementations in the IP library. The type inference problem is reduced to finding implementations from the library that can substitute the instances in the template. This process is called *IP selection*, which is discussed in [3], where we develop techniques that perform automated abstraction and design visualization-based matching of instances with real implementations from the IP library. Note that currently the selections are done at the structural level, and the next step would be to extend our CCL with behavioral specification and perform selection based on behavioral types [11].

The substitutions must not violate the existing type-bindings in the TC and must replace the type-variables in the TC by data types from the RTs of IPs. Therefore, we start by identifying instances of the architectural template with type-variable(s) in their type-vectors, so the algorithm can concentrate on resolving them with concrete types. A group of instances assigned common type-variables is called a *patch*. The objective is to substitute each instance with an implementation such that a patch can be correctly typed. Substituting every patch in the template with a set of implementations from the library would result in a solution for the type inference problem. We use a simple procedure to identify the patches in the template. In Figure 6.4, we illustrate patches that exist in the graphical visualization of a partially specified architectural template. The instances designated as **P** are partially typed and the others (**F**) are fully typed.

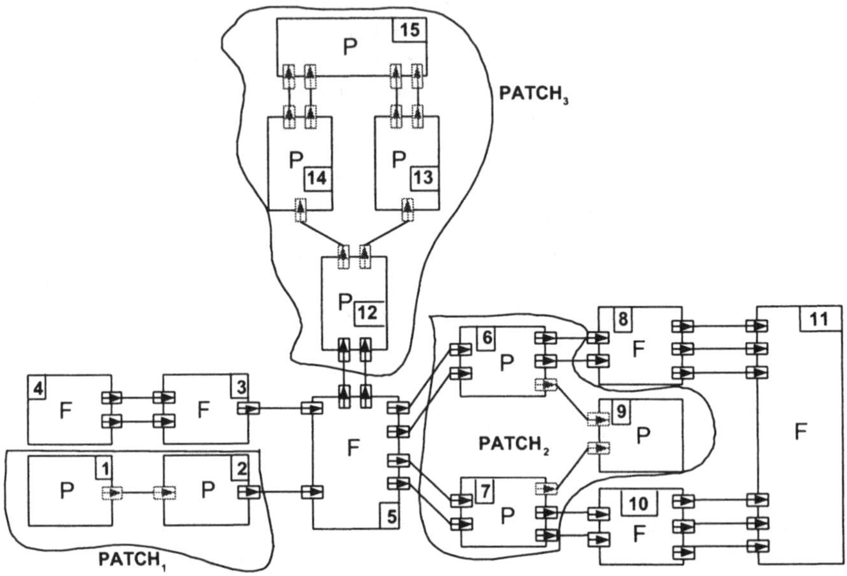

Figure 6.4 Illustration of patches in an abstract architectural template.

The patch identification algorithm is shown below. The procedure begins with an instance by setting the `curr_inst` variable and iterates through all the instances in I. The algorithm terminates when it has processed all the instances by marking them. During the processing of an instance, it is marked and either inserted as a part of the current PATCH being identified or else determined that it does not belong to any patch. An instance belongs to the current PATCH when it has at least one port that is type-constrained and the port's type-vector is assigned one or more type-variables, which is shared by at least one other instance in the current PATCH. An instance does not belong to the current PATCH, when it does not share its type-variables with any instance in the current PATCH. Finally, an instance does not belong to any PATCH when the instance is fully typed. The algorithm moves from one instance to another by traversing the port-to-port connections given by λ in TC and setting the `curr_inst` variable.

Initially, the PATCH is empty, and the procedure begins with some instance $i \in I$ (`curr_inst` = i). If i has a type-constrained port p, then i is marked and inserted in PATCH. The algorithm then proceeds to traverse the port-connections that i participates in using port p. These instances that interact with i through a port-connection that employs p are the targets for the next level of processing, where the procedure repeats by trying to identify whether they belong to the PATCH. A port-connection traversal comes to an end when either the target instance is already marked or has no type-constrained ports. All the instances collected in PATCH as a part of the traversals from i constitute a patch. Once all the traversals from the instance i has terminated, the PATCH is inserted into the PATCH_list and cleared to begin the next set of traversals.

O($max_i|\mu^{-1}(i)| \times |I|$) is the time complexity of the patch identification algorithm, where $max_i|\mu^{-1}(i)|$ is the maximum number of ports on any instance.

Algorithm 2: Patch Identification

{Given the TC = $\langle I, P, \mu_P, T, \gamma_P, \lambda \rangle$, which is output of Stage 1, a patch list PATCH_list = $\emptyset$, a instance list PATCH = $\emptyset$ and a variable `curr_inst` that points to the current instance being processed}

Step 1 For each $i \in I$,

 Step 1.1 `curr_inst` = i

 Step 1.2 If (!*marked*(`curr_inst`))

  then

   For each $p \in P$, **such that** $\mu(p)$ = `curr_inst`

    If (*type_constrained*(p))

    then

     mark(`curr_inst`);

     insert(PATCH, `curr_inst`);

     For each $c \in \lambda$, **such that** $c = (p, q)$ where $q \in P$

      `curr_inst` = $\mu(q)$;

      Repeat Step 1.2;

     End For_loop

   End For_loop

 Step 1.3 If (PATCH $\neq \emptyset$)

  then

   insert(PATCH_list, PATCH);

   clear(PATCH);

  else

   mark(`curr_inst`);

End For_loop

Application of Algorithm 2 on the architectural template in the example (shown in Figure 6.4) is shown in Table 6.1.

Table 6.1

Applying Algorithm 2

	`curr_inst`	**PATCH**	**Mark**	**PATCH_list**
I_1	I_1	I_1	I_1	$\emptyset$
-	I_2	$I_1;I_2$	I_2	$(I_1;I_2)$
I_2	I_2	$\emptyset$	-	$(I_1;I_2)$
I_3	I_3	$\emptyset$	I_3	$(I_1;I_2)$
I_4	I_4	$\emptyset$	I_4	$(I_1;I_2)$
I_5	I_5	$\emptyset$	I_5	$(I_1;I_2)$
I_6	I_6	I_6	I_6	$(I_1;I_2)$
-	I_9	$I_6;I_9$	I_9	$(I_1;I_2)$
-	I_7	$I_6;I_9;I_7$	I_7	$(I_1;I_2)$; $(I_6;I_9;I_7)$
I_7	I_7	$\emptyset$	-	$(I_1;I_2)$; $(I_6;I_9;I_7)$
I_8	I_8	$\emptyset$	I_8	$(I_1;I_2)$; $(I_6;I_9;I_7)$
I_9	I_9	$\emptyset$	-	$(I_1;I_2)$; $(I_6;I_9;I_7)$
I_{10}	I_{10}	$\emptyset$	I_{10}	$(I_1;I_2)$; $(I_6;I_9;I_7)$
I_{11}	I_{11}	$\emptyset$	I_{11}	$(I_1;I_2)$; $(I_6;I_9;I_7)$
I_{12}	I_{12}	I_{12}	I_{12}	$(I_1;I_2)$; $(I_6;I_9;I_7)$
-	I_{14}	$I_{12};I_{14}$	I_{14}	$(I_1;I_2)$; $(I_6;I_9;I_7)$
-	I_{15}	$I_{12};I_{14};I_{15}$	I_{15}	$(I_1;I_2)$; $(I_6;I_9;I_7)$
-	I_{13}	$I_{12};I_{14};I_{15};I_{13}$	I_{13}	$(I_1;I_2)$; $(I_6;I_9;I_7)$; $(I_{12};I_{14};I_{15};I_{13})$
I_{13}	I_{13}	$\emptyset$	-	$(I_1;I_2)$; $(I_6;I_9;I_7)$; $(I_{12};I_{14};I_{15};I_{13})$
I_{14}	I_{14}	$\emptyset$	-	$(I_1;I_2)$; $(I_6;I_9;I_7)$; $(I_{12};I_{14};I_{15};I_{13})$
I_{15}	I_{15}	$\emptyset$	-	$(I_1;I_2)$; $(I_6;I_9;I_7)$; $(I_{12};I_{14};I_{15};I_{13})$

The type substitution only needs to be concerned about patches, since the rest of the architectural template is fully typed. Therefore, the inputs to the type substitution algorithm shown below are the patches from the template (PATCH_list) and the RTs of implementations in the IP library. It also employs a data structure called the *config-record* (CR), which consists of three fields as shown here:

CR = { PT, MLIB : instance list;
`index` : instance;
}

The operation defined on the CR are *head, append, remove*, and *search*, which are list operations. Each iteration of the algorithm performs a select-substitute-check-save step, which modifies the CR using the defined operations. At the end of the iteration, the modified CR is saved onto a stack called the *CRstatus*.

Algorithm 3: Type Substitution Using IP Library

```
{Given PATCH_list, RTs from the IP library (LIB), the CR, and a CRstatus stack}
Step 1 Initialization of CR
       For each patch ∈ PATCH_list,
         For each i ∈ patch, append(PT, i);
         index = head(PT);
         MIB = ∅;
         CRstatus.push(CR);
Step 2 While(PT ≠ ∅ and !match-end)
         Perform select-substitute-check-save
         // Selection Phase
         1) For each RT ∈ LIB s.t RT ∉ MLIB
              Perform match(index, RT);
              If (successful)
              then
                append(MLIB, RT);
              Else
                match-end;
         // Substitute Phase
         2) Replace type-variables with data types;
         // Check Phase
         3) For each i ∈ search(PT)
              Perform match(i, LIB);
              If (successful)
              then
                remove(PT, i);
              Else
                Discard RT; Repeat Step 2;
         // Save Phase
         4) CRstatus.push(CR);
            index = head(PT);
            MLIB = ∅;
```

Step 3 If (PT = ∅)
then
type substitution is *complete*
Else If (*match-end* ∧ CRstatus = ∅)
then
failure
else
CR = CRstatus.pop(); Repeat **Step 2**.

During *selection*, the first instance from the PT list is matched with the RT of every implementation in the IP library using the function match(`index`, RT). This function matches an abstract port of the instance (pointed by `index`) with a concrete port of the selected implementation. The matching is successful with respect to that port if the data types of both the abstract and concrete port are equivalent or the abstract port's type is subsumed by the type of the concrete port. The question of "which abstract port to match with which concrete port?" is determined by the selection scheme chosen. If all the ports of the instance match with all ports of the selected implementation, then we have identified a possible implementation for the abstract instance. MCF employs various selection schemes that are described in [3], all of which are applicable for type substitution as long as they instantiate the type-variables and do not violate the existing type-bindings of the instance. A match instruments a fully typed specification of the partially typed instance, which happens during *substitution*.

At any point, if the selection fails to find a matching RT, then the algorithm has reached a *match-end*. This implies that the CR has an invalid configuration and is reverted to the previous configuration from the CRstatus, which helps the algorithm to continue. The algorithm terminates on two accounts: (1) PT list is empty and (2) reached a match-end and the CRstatus is empty.

The `index` field contains the instance that gets fully typed from the select-substitute-check-save iteration. The MLIB field collects the matched implementations during selection. The reason for maintaining this field is to restore the CR to a previous configuration and to enforce that a different implementation is selected in the next iteration. The *check* performs forward selections for certain instances in PT using the function match(i, LIB). These instances got fully typed as a result of the *substitution* apart from the `index`. They are identified using the search operation on PT. If the *check* results in a *match-end*, then the *select-substitute* phases are invalidated.

If the selected implementation is *overloaded*, then every data type allowable by the restrictions (from R) on the implementation is used to instantiate the type-variables until a match is found. If the selected implementation is polymorphic, then the type-variables are instantiated with $\sqcup$; the most generic type. The selection of an implementation is performed in the following order: fully typed $\rightarrow$ overloaded $\rightarrow$ polymorphic. If the second stage does not succeed in solving the type inference problem, then the given IP library is insufficient.

For the complexity analysis, let us consider the worst-case scenario, where the architectural template is one big patch. This implies that every virtual component in the template (I) is either untyped or partially typed. Now, let us assume that we have a library of N different IPs, where each has R type-restrictions specified on them. The time complexity of the algorithm is given by $O((N \times R)^{|I|})$, which is nonpolynomial in the size of the IP library ($N \times R$). This complexity is expected because, the algorithm takes an existential approach to the type inference problem, where it tries to see if there is at least one matching IP for every untyped/partially typed virtual component in the template.

Theorem 1. *The MCF type inference problem is NP-complete.*

`Proof:` Given a selection of RTs from the IP library, it is verifiable in polynomial time, that the substitution of partially typed virtual components in a TC by RTs will result in a fully typed TC. Hence the problem is clearly in NP.

For the NP-completeness proof, we reduce the problem of one-in-three monotone 3SAT [12] to the MCF type inference problem, similar to BALBOA [13] and LSE [14]. Monotone 3SAT is a problem of determining satisfiability of a logical expression with variables in conjunctive normal form, such that each clause in a set of disjunctive clauses is limited to three positive literals. One-in-three monotone 3SAT problem is used to determine whether there exists a truth assignment to the variables in the 3SAT, so that each clause has exactly one true literal.

The reduction proceeds as follows: Let us map the type `int` in the MCF type-system to the truth value "1" and all other legal MCF types to "0." Consider an instance F of monotone one-in-three 3SAT, with V variables and C disjunctive clauses. F is equivalent to a TC of an architectural template with C virtual components connected through V connections. The assignment of a truth value to the variable v of F is equivalent to the assignment of the corresponding MCF type to the ports of the connection v in the TC.

A clause c_a of F has three literals, namely l_{a_1}, l_{a_2}, and l_{a_3}, which are allowed exactly three possible assignments:

$$[1,0,0], [0,1,0], \text{ and } [0,0,1]$$

Each of these assignments is used to construct an RT, which belongs to the RT library for c_a as shown here:

$$\begin{aligned}RT_1 &= \langle\ \{m_1\}, \{p_1, p_2, p_3\}, \mu_P, \{0,1\}, \gamma_P, \emptyset, \alpha_P\ \rangle \text{ where,}\\ &\quad \gamma_P(p_1) = 1, \gamma_P(p_2) = 0, \text{ and } \gamma_P(p_3) = 0;\\ RT_2 &= \langle\ \{m_2\}, \{p_1, p_2, p_3\}, \mu_P, \{0,1\}, \gamma_P, \emptyset, \alpha_P\ \rangle \text{ where,}\\ &\quad \gamma_P(p_1) = 0, \gamma_P(p_2) = 1, \text{ and } \gamma_P(p_3) = 0;\\ RT_3 &= \langle\ \{m_3\}, \{p_1, p_2, p_3\}, \mu_P, \{0,1\}, \gamma_P, \emptyset, \alpha_P\ \rangle \text{ where,}\\ &\quad \gamma_P(p_1) = 0, \gamma_P(p_2) = 0, \text{ and } \gamma_P(p_3) = 1.\end{aligned}$$

Any of these assignments is legal **iff** a literal l_{a_1} in c_a, which is the same variable as the literal l_{b_1} in another clause c_b, has the same truth value as a result of the assignment. Therefore, the choice of RTs from the library of c_a and c_b must be such that the assigned truth values are the same for l_{a_1} and l_{b_1}. This is a version of the MCF type inference problem and hence the MCF type inference problem is at least as hard as the monotone one-in-three 3SAT, which is known to be NP-complete [12]. ■

6.3 COMPARATIVE STUDY

Our equality formulation is inspired by the type inference algorithm in ML [6], where the type equations are used to represent type constraints. Ptolemy II [15] follows an inequality formulation, because the type conversion performed using their type-system [16] naturally maps to inequality relations among types. In ML, the type constraints are generated from program constructs. In a component composition framework like MCF, where components are specified at the architectural-level devoid of an implementation detail, the framework does not have enough information about the component's behavior, so the components must provide their type information as either type-bindings to their ports, or by mapping undeclared ports to $\perp$. Similar to MCF, Ptolemy II requires its actors to either provide the type information on their declared ports or type constraints on their undeclared ports. However, during type inference in the first stage, MCF begins with $\perp$ on the undeclared ports of components in the architectural template and automatically creates type-bindings on the resolved ports and type constraints on the unresolved ones. This type inference is achieved using information of architectural connectivity as well as the static properties of the instantiated communication primitives (bus, switch, and so forth) in the template. The advantage is a rapid specification process allowing the modeler to create partially specified architectural templates, thereby shifting the complexity to the framework in inferring and constraining the template for IP composition.

The type resolution in Ptolemy II boils down to a problem of solving a set of inequalities defined over a finite type lattice [16]. The solution found by their algorithm is the most specific solution with lower implementation cost. The ML type inference algorithm finds the most general solution for a given program, which allows for maximal reuse. The type resolution in the stage 2 of MCF is a selection problem over a given IP library [3], which also finds the most specific solution. The solution set satisfying the constraints may not be a singleton, which is also true in the case of Ptolemy II. The most specific solution is the best matched IP intended by the modeler to solve the IP composition. Note that MCF can be employed to explore all possible solutions. However, a solution may not exist implying that the given IP library is insufficient to solve the problem. At this point, the type substitution algorithm is limited to selections based on type equivalence between an instance and the IP. As future work, we would employ subtyping relations to support type subsumption and type conversion, which would widen the solution space to solve the IP composition problem.

In the architectural template, a component is a placeholder for an IP, which could be fully typed, overloaded, or polymorphic. To capture them uniformly using the CCL, MCF allows for a partial specification, where the component has an incomplete TC. For an IP in the library, an RT is automatically synthesized from the IP [2] in MCF, where: (1) if fully typed ports, then the restriction set (R) is empty, (2) if overloaded ports, then the possible types are specified in R, and (3) if polymorphic ports, then the only type in R is $\sqcup$ (most general type). In standard ML, overloading of arithmetic operators must be resolved on appearance, but "eqtype" variables ranging over equality types are provided for the equality operator [6]. In Haskell, type classes are used to provide overloaded operations [7]. Ptolemy takes advantage of data encapsulation, where the tokens are active and know how to do arithmetic operations with another token.

6.4 CASE STUDY

We illustrate the construction of a module in a system model through IP composition. The module is the AMBA AHB bus. We start by creating the architectural template for the AMBA bus using the CCL in MCF. The bus consists of three masters and four slaves that communicate through a set of MUXes and DEMUXes that are driven by an arbiter and a decoder. The masters and slaves are described as instances

(L_{1-7}) of a leaf entity with three ports ($Rdata_i$, $Addr_i$, $Wdata_i$). The arbiter and decoder are also modeled as instances (L_8, L_9) of a leaf entity. The MUXes are specified as instances (M_{1-3}) of a merger entity, whereas the DEMUXes are instances (S_{1-3}) of a splitter. Two of the MUX (M_1, M_2) multiplex the requests from the masters and the third (M_3) multiplexes the acknowledgment from the slaves. These instances have external ports ($MASel, MRSel, MWSel$) driven through the arbiter ($ASel, RSel$) and decoder ($WSel$), respectively. The driver ports are shown as a doubles box on the instances. The bus template is shown in Figure 6.5.

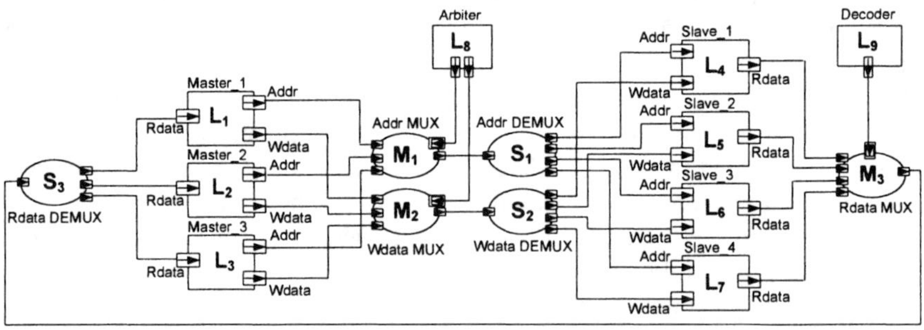

Figure 6.5 Architectural template of the AMBA AHB bus.

None of the instances have been characterized with type information and hence all ports are mapped to $\perp$. We begin by characterizing L_1 as follows:

$$[LRdata_1\text{:sc\_int}\langle 64\rangle,\ LAddr_1\text{:int},\ LWdata_1\text{:sc\_int}\langle 64\rangle]$$

The initial bus template with an incomplete TC shown below was created in under five minutes.

TC of AMBA AHB Bus Template

TC = {

I = $\{L_1, L_2, L_3, L_4, L_5, L_6, L_7, L_8, L_9, M_1, M_2, M_3, S_1, S_2, S_3\}$

P = $\{LRdata_1, LWdata_1, LAddr_1, LRdata_2, LWdata_2, LAddr_2, LRdata_3, LWdata_3, LAddr_3, LRdata_4, LWdata_4, LAddr_4, LRdata_5, LWdata_5, LAddr_5, LRdata_6, LWdata_6, LAddr_6, LRdata_7, LWdata_7, LAddr_7, ASel, RSel, WSel, MAddr, MAddr_1, MAddr_2, MAddr_3, MASel, MRdata, MRdata_1, MRdata_2, MRdata_3, MRSel, MWdata, MWdata_1, MWdata_2, MWdata_3, MWSel, SAddr, SAddr_1, SAddr_2, SAddr_3, SAddr_4, SRdata, SRdata_1, SRdata_2, SRdata_3, SRdata_4, SWdata, SWdata_1, SWdata_2, SWdata_3, SWdata_4\}$

$\gamma_P(LRdata_1)$ = sc_int⟨64⟩, $\gamma_P(LAddr_1)$ = int, $\gamma_P(LWdata_1)$ = sc_int⟨64⟩
$\forall$ p $\in$ P, if μ_P(p) $\neq$ (L_1) then γ_P(p) = $\perp$
$\lambda = \{(LAddr_1, MAddr_1), (LWdata_1, MWdata_1), \ldots\}$
}

Let us assume the type inference algorithm (first stage) begins with M_1. This results in binding all the multiplexing ports to *int* and constraining the driver port by the type-variable v_d. The algorithm exploits the property, which asserts that all multiplexing ports of a merger must have identical type-bindings. This maps to the following constraint:
if $\exists$ p_i, **s.t** $\gamma_P(p_i) \neq \perp$, **then** $\forall$ p_j, $(\gamma_P(p_j) = \gamma_P(p_i)) \vee (\gamma_P(p_j) = \perp)$, where p_i, p_j are multiplexing ports. The assignments performed by the algorithm for this iteration is shown here:

Step 1.1.1: - *association* based derivation
type-assignment($MAddr_1$, $LAddr_1$);
type-variable-assignment($ASel$, $MASel$);
$\forall i$, type-variable-assignment($LAddr_i$, $MAddr_i$), where 2 $\leq$ i $\leq$ 3;
Step 1.1.2: - *static-requirement* based derivation
type-assignment($MAddr$, $MAddr_1$);
$\forall j$, type-assignment($MAddr_j$, $MAddr_1$), where 2 $\leq$ j $\leq$ 3;
Step 1.1.1: - **new** derivations as a result of **Step 1.1.2**
type-assignment($SAddr$, $MAddr$);
$\forall k$, type-assignment($LAddr_k$, $MAddr_k$), where 2 $\leq$ k $\leq$ 3

Applying the above assignments result in the following type derivation for M_1, L_2, L_3, S_1, and L_8:

M_1: [$\perp$,$\perp$,$\perp$,$\perp$,$\perp$] $\rightarrow$ [v_a,int,v_b,v_c,v_d] $\rightarrow$ [int,int,int,int,v_d]
L_2: [$\perp$,$\perp$,$\perp$] $\rightarrow$ [$\perp$,v_b,$\perp$] $\rightarrow$ [$\perp$,int,$\perp$]
L_3: [$\perp$,$\perp$,$\perp$] $\rightarrow$ [$\perp$,v_c,$\perp$] $\rightarrow$ [$\perp$,int,$\perp$]
S_1: [$\perp$,$\perp$,$\perp$,$\perp$,$\perp$] $\rightarrow$ [v_a,$\perp$,$\perp$,$\perp$,$\perp$] $\rightarrow$ [int,$\perp$,$\perp$,$\perp$,$\perp$]
L_8: [$\perp$,$\perp$] $\rightarrow$ [v_d,$\perp$]

The algorithm is tuned toward picking a medium before a component as they are a good source for type derivations. Since M_1 associates with the splitter instance S_1 through ($MAddr$, $SAddr$), the next choice of the algorithm is S_1. It proceeds from $S_1 \rightarrow M_2 \rightarrow S_2 \rightarrow M_3 \rightarrow S_3 \rightarrow \ldots$ and terminates with the following configuration (final configuration):

L_i: [sc_int⟨64⟩, int, sc_int⟨64⟩], where 1 $\leq$ i $\leq$ 7,

L_8: [v_d,v_x],
L_9: [v_y]
M_1: [int,int,int,int,v_d],
M_2: [sc_int⟨64⟩,sc_int⟨64⟩,sc_int⟨64⟩,sc_int⟨64⟩,v_x]
M_3: [sc_int⟨64⟩,sc_int⟨64⟩,sc_int⟨64⟩,sc_int⟨64⟩,v_y]
S_1: [int,int,int,int,int]
S_2: [sc_int⟨64⟩,sc_int⟨64⟩,sc_int⟨64⟩,sc_int⟨64⟩,sc_int⟨64⟩]
S_3: [sc_int⟨64⟩,sc_int⟨64⟩,sc_int⟨64⟩,sc_int⟨64⟩,sc_int⟨64⟩]

The TC for the AMBA AHB bus template is incomplete at the end of the first stage. Therefore, we proceed to the next stage and perform type substitution. We start by identifying the patch of partially typed instances. There exist two such patches, namely $\{M_1, L_8, M_2\}$ and $\{M_3, L_9\}$. To illustrate the algorithm, we provide a oversimplified view of the IPs from the IP library relevant to the AMBA AHB bus.

RT of Implementations in the IP Library
RT_1 = { M = $\{C_1\}$, P = $\{p_1\}$, $\gamma_P(p_1)$ = sc_uint⟨2⟩, R = ∅ }
RT_2 = { M = $\{C_1\}$, P = $\{p_1,p_2\}$, $\gamma_P(p_1)$ = sc_uint⟨2⟩, $\gamma_P(p_2)$ = float, R = ∅ }
RT_3 = { M = $\{C_1\}$, P = $\{p_1,p_2\}$, $\gamma_P(p_1) = \gamma_P(p_2)$ = sc_uint⟨2⟩, R = ∅ }
RT_4 = { M = $\{C_1\}$, P = $\{p_1,p_2,p_3,p_4,p_5\}$, $\gamma_P(p_1) = \gamma_P(p_2) = \gamma_P(p_2) = \gamma_P(p_3) = \gamma_P(p_4) = \sqcup$, $\gamma_P(p_5)$ = sc_uint⟨2⟩, $\alpha_P(p_i) = r$ where $1 \leq i \leq 4$, R = $\{r\}$, where $r :=$ int | sc_int⟨64⟩ }

We demonstrate two different scenarios: (1) a straightforward type substitution for patch $\{M_3, L_9\}$, and (2) a type substitution through conflict resolution for patch $\{M_1, L_8, M_2\}$. The selection scheme used for the case study is data type-matching at the port level for an RTL implementation.

For patch $\{M_3, L_9\}$, the algorithm starts with the instance M_3. The only matching IP is that of overloaded RT_4 when the restriction r = sc_int⟨64⟩ is applied. This selection obeys all the type-bindings of M_3 and instantiates v_y with sc_uint⟨2⟩ during the substitution phase. The check phase finds the only other instance L_9 to also be fully typed as a result of the substitution. It then performs a forward selection and finds RT_1 to be a match for L_9, which completes the algorithm for this patch.

For patch $\{M_1, L_8, M_2\}$, the algorithm starts with M_1 and selects the IP with overloaded RT_2 (r = int) to be a match. During the substitution phase, the type-variable v_d is instantiated with sc_uint⟨2⟩. The following check phase does not find any other fully typed instance in the patch and the current configuration is saved on the stack. The algorithm proceeds with L_8 and selects RT_2 as a possible match, which instantiates the type-variable v_x with float. In the check phase, M_2 is found to be fully typed from the substitution on L_8 and subjected to forward selection. However, no IP is found that matches with M_2 and this results in a match-end. This is a conflict that implies the selection resulted in a configuration on which the type inference fails. Therefore, we discard the selection on L_8 and search for a new match. This results in selecting RT_3 as a match for L_8, which instantiates v_x with sc_uint$\langle 2 \rangle$. During the forward selection in the check phase, M_2 is found to be matching with overloaded RT_4 (r = sc_int$\langle 64 \rangle$) and this successfully completes the patch. The complete TC is used to plug-in the architectural template with IPs and create the executable model.

6.5 SUMMARY

Important issues in component-based system-level design environments used to create high-level models are how to reduce design time and allow for rapid DSE. One approach is to use a CCF that allows designers to architect the system in terms of components and their composition and through automated type inference-based composition of IP-cores, create simulation models for exploring various design alternatives. This chapter addresses how to simplify strong typing requirements imposed by C++-based methodologies through an architectural specification in MCF that allows for flexible components and an IP library. We have formulated the type inference problem that exists in MCF and propose a multistage solution that employs type-assignment and type substitution. Our future work will also include behavioral typing to see how compatible two IPs are, and what is needed to substitute one for the other.

References

[1] D. A. Mathaikutty and S. K. Shukla, " A Metamodel for Component Composition," FERMAT Group, Virginia Tech, Tech. Rep. 2005-13, December 2005.

[2] D. A. Mathaikutty and S. K. Shukla, "Mining Metadata for Composability of IPs from SystemC IP Library," *Proceedings of Forum on Specification and Design Languages (FDL'06)*, TU Darmstadt, September 2006.

[3] D. A. Mathaikutty and S. K. Shukla, "SoC Design Space Exploration Through Automated IP Selection from SystemC IP Library," *Proceedings of IEEE International SoC Conference (SOCC'06)*, Austin, TX, September 2006, pp. 109–110.

[4] FERMAT Group, MCF Web site, http://fermat.ece.vt.edu/MCF/MCF.htm, 2005.

[5] W. Putzke-Roming, M. Radetzki, and W. Nebel, "Objective VHDL: Hardware Reuse by Means of Object Oriented Modeling," *Proceedings of the 1st Workshop on Reuse Techniques for VLSI Design*, September 1998.

[6] R. Milner, "A Theory of Type Polymorphism in Programming," *Journal of Computer and System Sciences*, Vol. 17, 1978, pp. 348–375.

[7] P. Wadler and S. Blott, "How to Make Ad-Hoc Polymorphism Less Ad-Hoc," *16th ACM SIGPLAN-SIGACT Symposium on Principles of Programming Languages*, 1989, pp. 60–76.

[8] J. Warmer and A. Kleppe, *The Object Constraint Language: Getting Your Models Ready for MDA*, 2nd Edition, Reading, MA: Addison-Wesley, 2005.

[9] OSCI Group, SystemC Web site, http://www.systemc.org/.

[10] Spirit Consortium, "SPIRIT Schema Working Group," http://www.spiritconsortium.org/.

[11] L. Cardelli, "Type Systems," *ACM Computing Surveys*, Vol. 28, No. 1, 1996, pp. 263–264.

[12] M. Garey and D. Johnson, *Computers and Intractability: A Guide to the Theory of NP-Completeness*, New York: WH Freeman & Co., 1979.

[13] F. Doucet, S. Shukla, M. Otsuka, and R. Gupta, "An Environment for Dynamic Component Composition for Efficient Co-Design," *Proceedings of Design Automation and Test Conference Europe (DATE'02)*, 2002.

[14] M. Vachharajani, N. Vachharajani, S. Malik, and D. August, "Facilitating Reuse in Hardware Models with Enhanced Type Inference," *Proceedings of the IEEE/ACM/IFIP 2nd International Conference on Hardware/Software Codesign and System Synthesis (CODES+ISSS'04)*, September 2004, pp. 86–91.

[15] E. A. Lee, "The Ptolemy Project," http://ptolemy.eecs.berkeley.edu/.

[16] Y. Xiong, "An Extensible Type System for Component Based Design," University of California Berkeley, Electrical Engineering and Computer Sciences, Ph.D. Thesis, 2002.

Chapter 7

IP Composition

Consider the following problem confronted by a designer: Given a library of C++-based IP blocks that may contain implementations of various standard blocks needed for a system level model, the designer has to create a system-level model for solving a problem P by combining the IP blocks selected from the given library. There may be multiple outcomes for this exercise by the designer: (1) there are not enough required and relevant IP blocks in the library to build a system model for solving the given problem P, (2) there is a way to combine the IPs but it requires programming glue to combine these IPs, and (3) there are multiple possible ways to combine the given IPs to create the model and the designer has to choose one.

Let us assume that the designer has a tool to do this without having to worry about C++ programming and software engineering issues. Instead the tool facilitates visually creating a module/connection/bus-based microarchitecture. Moreover, the tool does the exploration of the IP library to instantiate the microarchitecture with appropriate real implementations for the designer. What if the designer has such a tool that also declares the impossibility of the exercise or helps with the DSE, as the case may be?

A recent industrial trend in system level design is to use high-level programming languages such as C++ to build custom design exploration and analysis models, especially with the increased usage of SystemC. The proliferation of SystemC IP cores necessitates design tools and methodologies that reuse IPs from an IP library for construction of system models. SPIRIT [1] proposes an approach to reusing IP cores to enable IP configuration and integration by utilizing metadata exchanged in a design-language neutral format between tools and SoC design flows. However, the designer is required to specify the IP-explicit metainformation, which is a time-consuming task and affects the essential rapid DSE. Furthermore, no tool such as the one envisioned above has been developed by SPIRIT.

Our tool is a component composition framework (CCF) that allows designers to: (1) describe the components and their interactions with a semantically rich visual front-end, (2) check for system-level inconsistencies, type mismatches, and interface incompatibilities enforced during modeling, (3) perform automated selection and composition of IPs based on structural type theoretic principles from given IP libraries, and (4) generate executable models and perform validation of functionality.

A CCF provides the language for such component compositions (CCs) and automatically selects and substitutes the abstract components by matching implementations from a library of IP cores. The end result is multiple executable models to analyze performance, simulation efficiency, and so forth, which enables rapid DSE. We provide a new capability in our CCF such that the IP libraries can be "generic," in particular, if SystemC IP library is considered, the IPs can be templated, and hence may represent multiple IPs with single library elements. The challenges in IP composition using a generic library is discussed in this chapter. We also describe the "role" and "responsibility" of a CCF user that we call the *interaction pattern* and discuss the automation issues with our CCF at the interaction level. Currently, our structural type based approach proposes a solution to the problem with a limited guarantee of correctness.

We developed MCF, a metamodeling-based component composition framework for SystemC based IP core composition. It is built on top of GME [2] and captures the syntax and semantics of the visual composition language into a metamodel [3]. The metamodel is developed using Unified Modeling Language (UML) and Object Constraint Language (OCL). The design environment allows a modeler to visually construct abstract architectural templates conforming to the underlying semantics of the metamodel. The library consists of compiled SystemC IPs on which MCF performs reflection [4] to extract metadata in XML format and creates an encapsulating manager that allows introspection of the metadata. Furthermore, an IP restriction technique has been added that captures constraints on generic IPs in a database and has a manager that allows querying the database. MCF provides a tool for automated selection that interacts with the IP library and constraint database through the respective manager to perform selection. Finally, an interaction console ties in the different ingredients and techniques to enable CC using MCF.

We also show the addition of an ingredient called *IP restricter* to allow generic IPs within our library. We also discuss the task-flow necessary for IP composition and the roles played by a user employing the MCF and the interaction perspective in these roles. Details of the type inference, use of typing rules for automated IP selection, and the modeling framework are detailed in the other chapters as well as in [3, 5, 6]. The overarching principles of MCF are the focus of this chapter.

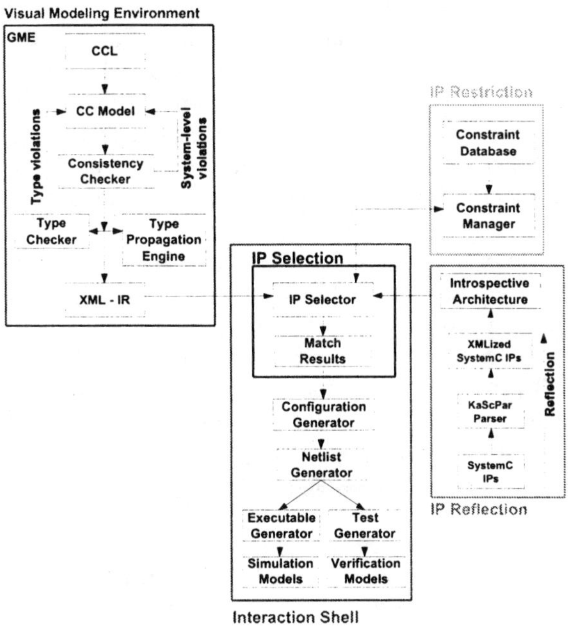

Figure 7.1 Design flow diagram.

Design Flow

The designer begins by creating a component composition model, which is an architectural specification of the system using the Component Composition Language (CCL). The designer ends with various simulation and verification models for the system design. The abstract architectural template of the system to the various executables is enabled through our CCF design flow shown in Figure 7.1. As stated before, it consists of five main ingredients: (1) visual modeling environment, (2) IP library with reflection and introspection capability, (3) IP restricter, (4) IP selector, and (5) interaction console.

The modeling environment facilitates the description of an abstract model of the system. Upon automated checking of conformance to the various constraints and completion of the model, it is converted to an intermediate representation (IR) in XML format. We have a library of compiled objects of C++ IPs from which the composition-related metadata is extracted through our automated reflective mechanism into XML format. To allow the exchange of the reflected metadata between the ingredients, we develop an introspective architecture that allows for querying. A constraint database captures the type-restrictions on generic IPs and a constraint manager provides a querying technique for the other ingredients. Using the interaction shell, the IP selector is given the IR of the abstract model, the XMLized library, and the constraint database as input to create executable and verification models.

The two types of users of the MCF are *modelers* and *library engineers*. The tasks of a modeler are as follows: (1) describe a CC model using the modeling framework, (2) perform consistency and type checks, (3) perform type propagation, (4) generate an IR for the CC in XML, (5) perform IP selection, (6) configure the selection, (7) perform composition through netlist creation, (8) generate an executable specification, and (9) perform validation through testbench creation. Most of these tasks are automated, which implies that the modeler is only required to click buttons or type commands on a console.

The tasks of a library engineer are as follows: (1) implement different IP cores, (2) populate the IP Library, and (3) specify restrictions on generic IPs. A library engineer may play the role of a modeler in which case we have an "expert user." If a user has insight into IPs in the library such as naming conventions, and levels of abstraction, then navigating MCF to create executables can be accelerated.

7.1 MCF INGREDIENTS

We briefly outline the contributions of the existing ingredients of MCF shown in Figure 7.1.

Ingredient 1 - Visual Modeling Environment: Our modeling environment built on top of GME provides a visual language for describing component composition models (CCMs). A CCM is an abstract architectural template of a system in terms of components and their composition. The entities allowed in a CCM are *components* and *media* at varying abstraction levels such as RTL and transaction level (TL) with hierarchical descriptions. The visual language also enforces structural composition rules that describe the possible interconnections between the different entities that a modeler can instantiate and characterize to describe the system. The specifics of ingredient 1 are provided in Chapter 4.

Ingredient 2 - IP Library: Our library is a collection of compiled SystemC IP cores from the SystemC distribution [7] and industry contribution. Given a library to perform IP selection, composition, and validation, we need to reflect composition related metadata and create an introspective architecture. In the following section, we outline our R-I capability.

Ingredient 2.1 - IP R-I: The IP reflector extracts two kinds of metadata; one related to SystemC components and the other related to SystemC channels. The metadata on an RTL component contains the information on input-output ports and clock ports. For these ports, information related to the datatype and bitwidth are essential constituents of the metadata. For a component/channel at TL, it contains information on interface ports, clock ports, and interface access methods. Therefore the metadata would contain the function signatures that embed the function arguments and return types. For these arguments and return types, the datatype and bitwidth need to be correctly extracted. The hierarchical structure of an IP is also extracted as a part of metadata. The details on the composition-related metadata extraction from SystemC is provided in [4]. The extracted information in captured in a data structure, which is an XML document model object. The appropriate C++ APIs have been implemented that allow introspection of the XML data structure.

Ingredient 3 - IP Selector: The primary task of the MCF is to select an appropriate IP from the library that can be plugged into the architectural template of the system to create possible executable models that allow for rapid exploration of design alternatives. We discuss the IP selection problem in MCF and propose techniques that perform automated abstraction and design visualization-based matching from an abstract architectural template of the system in [5].

The main contributions of the IP selector are the different selection schemes, which are summarized as follows. (1) For quick selections, we search based on nomenclature (version number, IP name, library name, and distributor). (2) For RTL components, we search based on port level structural type (data types). (3) For TL components and channels, we search based on interface-level structural type (function signatures and interface port specifics). (4) Searching based on visualization of the IP, namely black-box, flattened, and hierarchical. (5) Mix (1), (2), and (3) with opacity in (4) to obtain nine different selection schemes. (6) An exhaustive search using the nine selection schemes to obtain the ideal match. The selection schemes take advantage of the composition related metadata provided by the R-I capability of the IP library. The specifics of ingredients 2 and 3 are provided in Chapters 5 and 6.

7.2 HANDLING GENERIC IPS

In order to promote reuse, we must allow for generic IPs in the library such as FIFOs, memories, and switches that are polymorphic and are implemented using C++ templates in SystemC (example shown in Figure 5.8). In order to incorporate these in our library, the reflection process should provide the polymorphic parameters of these IP implementations and the corresponding ports and interfaces dependent on these parameters.

These IPs promote reusability through context independent usage, but not all of these are truly generic. We define true genericity to be the case when the implementation is neither bound by the way it communicates, nor by the type of computation[1] it performs. This is true for FIFOs and memories as they act as containers for values of a homogeneous type. However in hardware description, it is common to have implementations that are polymorphic with respect to a finite set of type instantiations. Consider the example of a polymorphic adder, which is restricted to input types ranging from integer to floating point types of varying bitwidth. It is unlikely to have a polymorphic adder that performs the above as well as matrix or complex number summations. This restriction is imposed by how the adder has been implemented and we refer to these as *partially generic*. Currently in C++ there is no way of capturing these restrictions, which are essential to avoid run-time errors. Therefore, to restrict a polymorphic description to a set of allowed types, we implement *IP restriction*.

For polymorphic IPs, we are provided with a constraint file, which describes type restrictions on the generic aspects of the implementation. The specification process is based on an XML schema called *CXML* that has two parts, first the type restrictions, and the second part that enforces these restrictions on various IPs as shown in Figure 7.2.

Given the constraint file, we construct a constraint graph that serves as a data structure. As part of restricting IPs, we also develop an introspective manager that provides a querying mechanism.

1. Uniformly applies the computation on any type

```
<constraint_db name = "CDB">
<constraint_entry name = "C1" index = "1" >
<dt_entry type = "int" length = "0"/>
<dt_entry type = "sc_int" length = "32"/>
<dt_entry type = "sc_uint" length = "64"/>
</constraint_entry>
...
<component name = "fifo">
<c_entry id = "C1"/>
</component>
...
</constraint_db>
```

Figure 7.2 Constrained FIFO.

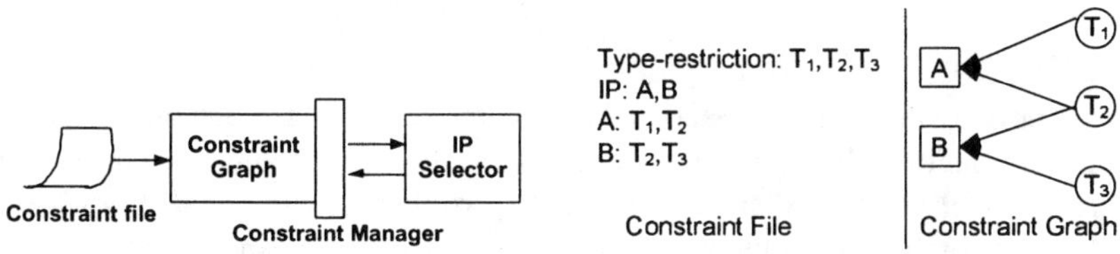

(a) Design flow (b) CG example

Figure 7.3 (a, b) IP restriction.

Our design flow shown in Figure 7.3(a) allows the library engineer to specify type-restrictions and update them as the implementation changes. It also allows specifying IP independent type restrictions that can be enforced on multiple polymorphic IPs. The mapping between a type-restriction and an IP are done via positional value, which for an IP is the position of the template argument and for a *constraint_entry* is the *index* attribute of its items. The type-restrictions are extracted from the constraint file and used to populate a structure called the constraint graph (CG). It captures the relationship between the generic IPs and the type-restrictions (Figure 7.3(b)) and can be seen as a constraint database (CDB). It is queried during IP selection to identify the type-restrictions on the IP being matched. In order to facilitate the Q and A, we have an introspective manager that plays middle ground between the IP selector and constraint graph. The questions allowed through the introspective manager are as follows: (1) does an IP have any type-restriction enforced on it, (2) for any given IP, what type-restrictions are enforced, and (3) for any given restriction on an IP, what is the set of legal types?

7.3 INTERACTION PATTERN

We describe the interactions of a modeler and a library engineer with the CCF.

Modeler: The interaction of a modeler has three parts: (1) modeling activity, (2) IP selection, and (3) generation of the executable models. The third part of the interaction is highly dependent on whether the CCF can resolve the CC problem. The CC problem is stated as:

Definition 17. ***CC Problem***
Given an architectural template of the system and an IP library, can MCF select from the IP library to match against abstract components in the template and compose them to create executable models?

Applying the CCF approach to this problem results in the outcomes discussed in the introduction. If the CCF fails to resolve the CC problem, then either the abstract model is not sufficient to solve the problem or the relevant IPs are not available in the library. In the first case, the modeler can revisit the abstract model to characterize it with more system-specific details and reattempts the CC problem through MCF. In the second case, MCF informs the modeler that with its current resources a solution was not found.

For this outcome, the modeler performs the first two parts of the interaction. The first part is bound to the visual modeling environment, where not just the abstract template is created but the modeler has to conform to the various design-time checks and create the intermediate representation. These tasks are automated through different analysis engines that are integrated into the modeling environment. The user is only required to click the different buttons and fix the violations flagged during the modeling process. In the second part of this interaction, the selection process is initiated through the interaction shell. The modeler specifies the selection code that informs the shell as to which selection criteria should be used to search for matching IPs. At the end of the interaction, the match results are generated that would be empty if the selection procedure cannot find any matching IPs or partial if the matching failed for some.

The other possible outcomes are that one/many solutions for the CC problem exists from the given IP library. In these cases, the modeler performs all three parts of the interaction. The third part of the interaction is also enabled through the shell, which results in simulation or verification models for the abstract specification. This part of the interaction enables the CCF to bind the abstract components with abstract ports to the actual implementation with concrete ports. It cannot be completely automated, since it is closely tied to the criterion chosen during selection. For example, if a name-based selection was performed, then components and implementations are matched on their naming conventions, but this does not mandate the internal ports or interfaces of the two to have similar naming conventions. In the Adder example shown in Figure 7.4, the name-based selection succeeds, but the name-based automatic port binding fails. Therefore, the port binding requires the user to manually bind the ports.

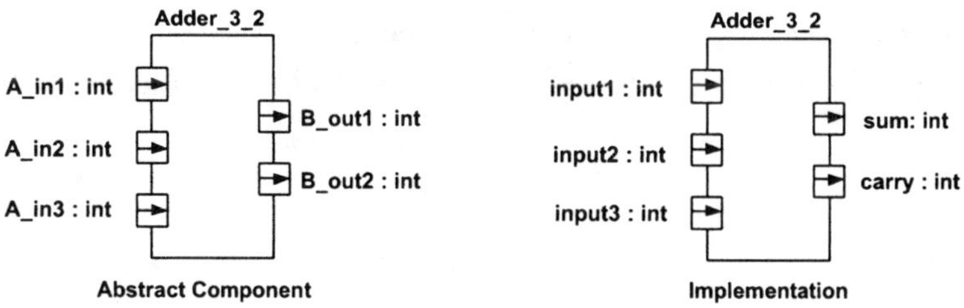

Figure 7.4 Adder example.

Similar issues exist for the other matching criteria because the selection is performed based on different compositional characteristics and remains devoid of any implementation specifics. However, we assume the best case and perform an automatic binding. In our future work, we plan to address this problem by establishing a naming standard that enforces a common way of specifying ports and interfaces. Once the binding process completes, the simulation model and testbench generation are performed automatically.

For the case with multiple solutions, the third part of the interaction deals with a larger solution set. As a result, a configuration phase is added to sift through the solution set. Currently, the user is required to configure the result set and tag the *ideal match*. In the future, we would like to automatically determine the most ideal match based on some performance numbers on the matched IPs, if the IPs are precharacterized. On completion of the configuration, the interaction facilitates CCF binding and automatic executable generation.

Library Engineer: Here the interaction is with SystemC to develop the IP cores and XML to specify the constraint file with type-restrictions on generic IPs.

7.3.1 Interaction Console

The task-flow through the C++-based interaction shell is divided into three stages: *initialization*, *selection*, and *executable generation*. The shell commands that make this interaction possible are shown in Figure 7.5.

```
Usage Commands:
read_ddb <filename>               [ Populate the IP database ]
read_ccm [opt] <file>             [ Populate the CCM DOM ]
read_cdb <filename>               [ Populate the constraint database ]
perform selection                 [ Perform a full blown selection ]
perform nselection                [ Perform a name-based selection ]
perform fselection                [ Perform a flattened selection ]
perform hselection                [ Perform a hierarchical selection ]
perform RTLselection              [ Perform an RTL-level selection ]
perform TLselection               [ Perform a TL-level selection ]
gen_config                        [ Generates the configuration ]
gen_netlist                       [ Generates the netlist ]
gen_toplevel                      [ Generates sc\_main module ]
gen_tb                            [ Generates testbench ]
print_match                       [ Generates match results ]
result_log                        [ Generates a log file ]
debug_log                         [ Generates a log file ]
done                              [ Process complete ]

operators:
        +                         [ Concatenate two selection cmds ]
```

Figure 7.5 Shell commands.

Initialization Stage: The user loads the respective data structures with the IR of the CCM, the XML IP metadata, and the CDB, using the commands **read_ddb**, **read_ccm**, and **read_cdb**.

Selection Stage: We allow both single-mode selection (such as **nselection**) as well as mixed mode selection using the "+" operator (such as **nselection + fselection**). Internally, these commands are mapped to a set of selection codes that guide the IP selector. We also have the option for a full-blown selection, where the user does not inform the IP selector on how to perform the selection. In this mode, the IP selector exhaustively applies all the searches to identify a possible match. In this case, the search begins with the most restricted criteria and if a match is not found, then the selector looks for more generic search patterns. The match result from the selection process is visualized using the **print_match** command. The shell also provides commands such as **result_log** and **debug_log** to view and debug the selection results.

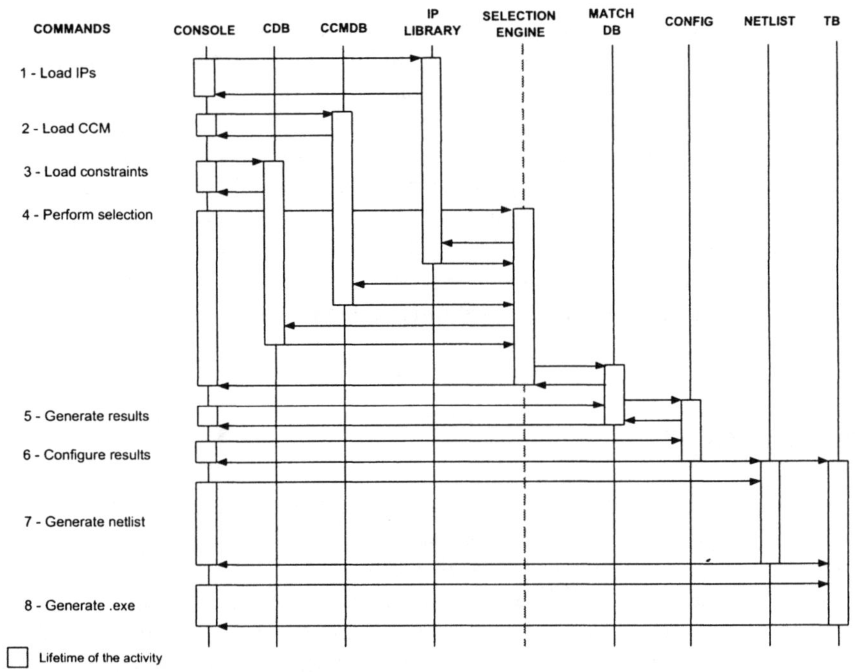

Figure 7.6 Task flow diagram.

Executable Generation Stage: In order to generate executables for the CCM, the selection stage should have succeeded (outcome 2 and 3). In case of outcome 3, the match results need to be configured, where one of the many matching IPs will be picked and configured as the *ideal match*. The **gen_config** command allows the modeler to look through the search results and inform the shell as to which one to tag as the *ideal match*. The next set of tasks are binding and generate executables.

If a name-based selection was performed, then the task uses the naming conventions to bind the ports and interfaces. Similarly, if RTL or TL selection was performed, then the task uses port-level datatypes and interface-level function signatures to bind ports. We provide automated binding (**gen_netlist**) during which we try to bind the ports and interfaces by looking at the selection criterion; however, often this fails and requires manual binding.

Simulation or verification model generation is the final task of this stage. For simulation models, the shell prints out an *sc_main* module in SystemC with all the implementation blocks hooked up through signals (**gen_toplevel**). The procedure for creating the toplevel module takes into account the abstraction level of the IPs and is very specific to SystemC; hence not discussed here. The verification model generator creates SCV-based constrained and unconstrained randomized testbenches for the SystemC model (**gen_tb**). Our work on test generation is detailed in [8]. The selection and testbench generation are performed through the usage of a common data structure, which serves as the backbone for all the processing enabled through the shell. The data structure facilitates a scalable user interaction, where multiple abstract descriptions are processed at the same time and different selections are performed on a given abstract description.

The complete task flow enabled through the interaction console is shown in Figure 7.6.

7.4 CASE STUDY

Consider the scenario where an audio filter for a digital equalizer is being designed. The modeler wants to reuse an FIR core from the given IP library to design the audio filter and proceeds by creating an architectural template of the core. The modeler visualizes the core as an encapsulation of a controller and a convolution block and creates the CCM shown in Figure 7.7. The stimulus and display are placeholders for components that would test the FIR IP after selection and composition. The main block in the CCM is the **fir_computation** that is modeled as a hierarchical component with four inputs and two outputs (Figure 7.7(a)). The inputs are *CLK*, *CLEAR*, *IN_READY*, and *DATA* and the outputs are *OUT_READY* and *VALUE*. These are modeled as hports (hierarchical ports) and cports (clock ports) and specify the structural interface of the IP. It also contains two leaf-level blocks, namely **controller** and **convolutor** modeled as shown in Figure 7.7(b).

Type bindings are provided for each of the blocks, which is essential to correctly identifying an IP that is structurally compatible with the CCM description. The type binding specified for the **fir_computation** block is shown here:

```
TB_fir_computation: { (cport_i, CLK, bool),
(hport_i, CLEAR, bool), (hport_i, IN_READY, bool),
(hport_i, DATA, float), (hport_o, OUT_READY, bool),
```

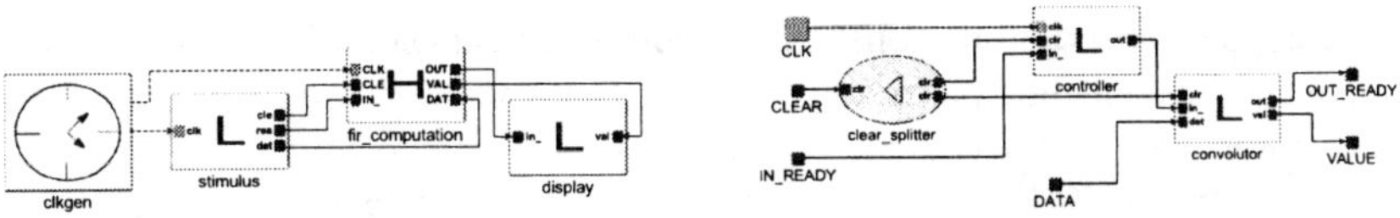

(a) FIR toplevel. (b) **fir_computation** block.

Figure 7.7 (a, b) Component composition model of FIR filter.

```
(hport_o, VALUE, float), TB_controller, TB_convolutor }
```

Note that the type bindings of the controller and convolutor are part of the type binding for the **fir_computation** block, which preserves the structural hierarchy. Furthermore in Figure 7.7, the blocks do not have behavioral annotations, but are described structurally (port specifics) and hence seen as an abstract architectural template. Now consider the RTL FIR filter in the SystemC distribution [7], which is modified and made generic with respect to the input sample and convolution result for the purpose of the following illustration. It consists of a toplevel component, namely **fir_top**, which the library engineer modifies to make it polymorphic to the sample type. The *fir_top* is hierarchical and embeds the **fir_data** and **fir_fsm** components. This IP is subjected to reflection and the metadata obtained is shown here:

```
component: fir_top<T>
input  = { RESET: bool, CLK: bool, IN_VALID: bool,
           SAMPLE: T }
output = { OUT_DATA_RDY: bool, RESULT: T }
subcomponent: fir_fsm
input = { reset: bool, clock: bool, input_valid: bool }
output = { state_out: bool }
subcomponent: fir_data<T>
input = {reset: bool, state_out: bool, sample: T }
output = {out_data_rdy: bool, result: T }
connection(c1, RESET, reset)
connection(c2, CLK, clock)
...
```

Note that input data and output result exchanged through the IP ports *sample*, *SAMPLE*, *result*, and *RESULT* are of type T, which the library engineer needs to constraint to avoid unnecessary selections that result from an IP being generic.

Therefore, the constraint file consists of type-restrictions on the **fir_top** and **fir_data** components as shown here:

```
<constraint_db name = "CDB">
<constraint_entry name = "T1" index = "1" >
<dt_entry type = "int" length = "0"/>
<dt_entry type = "sc_uint" length = "64"/>
</constraint_entry>
<constraint_entry name = "T2" index = "1" >
<dt_entry type = "float" length = "0"/>
</constraint_entry>
...
<component name = "fir_top">
<c_entry id = "T2"/>
</component>
<component name = "fir_data">
<c_entry id = "T1"/>
<c_entry id = "T2"/>
</component>
...
</constraint_db>
```

In order for MCF to select the IP **fir_top** as a pluggable instance for the abstract **fir_computation**, the correct selection criteria must be identified that would enable the framework to match the abstract component and the concrete implementation. The CCM is described at the RTL abstraction and the **fir_computation** is hierarchical, therefore the best criterion to obtain an ideal match would be (**RTLselection** + **hselection**). This mix-mode criterion would allow the framework to visualize the **fir_top** as shown in Figure 7.8 and proceed by matching each *hport*, *cport*, and *lport* (leaf port) in the type binding of the **fir_computation** with the **fir_top**'s concrete input and output ports.

A match at the port level is found, when the data types are similar or are related through type subsumption [6]. A match at the component level is found, when all the ports as well as embeddings match. While matching with a polymorphic IP, the framework queries the CDB to search for type-restrictions specified on the IP and propagates the type information to constrain the IP. In this case study, the framework chooses the type-restriction **T2** that constrains both **fir_top** and **fir_data** to obtain a match as shown here:

```
-------- BEGIN MATCH --------------------------
```

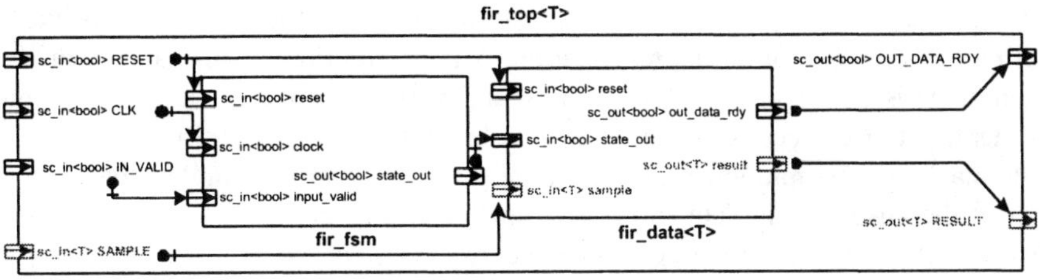

Figure 7.8 Hierarchical view of **fir_top**.

```
//fir_computation Vs fir_top
2 bool hport_i + 1 bool cport_i = 3 bool input
1 float hport_i = 1 float input
1 bool hport_o = 1 bool output
1 float hport_o = 1 float output
//controllor Vs fir_fsm
1 bool lport_i + 1 bool cport_i = 2 bool input
1 bool lport_i = 1 bool input
1 bool lport_o = 1 bool output
//convolutor Vs fir_data
2 bool lport_i = 2 bool input
1 float lport_i = 1 float input
1 bool lport_o = 1 bool output
1 float lport_o = 1 float output
-------- MATCH FOUND --------------------------
```

If more than one match is found, then the modeler is allowed to pick the one that seems an ideal match. For the **stimulus** and **display** blocks, the framework can be used to select an input generator and a monitor from the library if they exist, or can be automatically synthesized. The matched IPs are plugged into the abstract template to generate a netlist. An important part of netlist generation is the binding of the abstract and the concrete ports. Currently, we require hints from the modeler to identify the correct implementation of an abstract port. Our future work will automate the process using a port naming convention from the standards provided by SPIRIT [1]. Finally the executable generation, which in SystemC maps to creating the toplevel using *sc_main* or wrapping these IPs in an *sc_module*, is performed automatically by the framework.

A limitation of the framework is that the behavior of an IP is not considered during the selection of an implementation. A large solution set of matched implementations is obtained from a pure structural search and the modeler is required to identify the correct subset of matches. In future work, we will capture finite state machine or communicating sequential process style behavioral annotations as part of the abstract template and reflect more than just compositional metadata to facilitate IP composition with guaranteed correctness.

7.5 SUMMARY

Important issues in component-based system-level design environments used to create high-level models are how to reduce design time and allow for rapid DSE. Our approach is to use a CCF that allows designers to architect the system in terms of components and their composition, and through automated type inference-based selection and composition of IP cores, create simulation models for exploring various design alternatives. This chapter introduced the main principle and the ingredients that facilitate the above objective. It also discussed how we handle generic implementations within our IP library. Furthermore, we described the possible interactions that the different users are allowed through the CCF.

References

[1] Spirit Consortium, "SPIRIT Schema Working Group," http://www.spiritconsortium.org/.

[2] A. Ledeczi, M. Maroti, A. Bakay, G. Karsai, J. Garrett, C. Thomason, G. Nordstrom, J. Sprinkle, and P. Volgyesi, "The Generic Modeling Environment," *Proceedings of Workshop on Intelligent Signal Processing* Budapest, Hungary, May 2001.

[3] D. Mathaikutty and S. Shukla, "A Metamodel for Component Composition," FERMAT Lab., Virginia Tech, Tech. Rep. 2005-13, December 2005.

[4] D. Mathaikutty and S. K. Shukla, "Mining Metadata for Composability of IPs from SystemC IP Library, " *Design Automation for Embedded Systems*, Vol. 12, No. 1-2, 2008, pp. 63–94.

[5] D. A. Mathaikutty and S. K. Shukla, "SoC Design Space Exploration Through Automated IP Selection from SystemC IP Library," *Proceedings of IEEE International SoC Conference (SOCC'06)*, Austin, TX, September 2006, pp. 109–110.

[6] D. Mathaikutty and S. K. Shukla, "Type Inference for Component Composition," *Proceedings of 5th ACM-IEEE International Conference on Formal Methods and Models for Codesign (MEM-OCODE'07)*, May 2007, pp. 61–70.

[7] OSCI Group, SystemC Web site, http://www.systemc.org/.

[8] H. D. Patel, D. A. Mathaikutty, D. Berner, and S. K. Shukla, "CARH: A Service Oriented Architecture for Validating System Level Designs," *IEEE Transactions on Computer-Aided Design of Integrated Circuits and Systems*, Vol. 25, No. 8, August 2006, pp. 1458–1474.

Chapter 8

Checker Generation for IP Verification

In the previous chapters, we presented the theory, techniques, and ingredients enabling MCF to perform IP composition. MCF utilizes selection techniques [1] and structural characteristics (IP metadata [2]) to provide a solution to the IP-reuse problem. This structural approach to IP selection only provides a limited guarantee of correctness. The notion of correctness is that the virtual component and the IP implementation are equivalent with respect to their skeletons and embeddings. In this chapter, we boost our MCF design flow presented in Chapter 7 (Figure 7.1), with newer ingredients, in order to enhance the guarantee of correctness for our CCF-based solution to the IP composition problem. The additions to MCF are property specification and checker synthesis, which enable a behavioral approach to IP composition.

8.1 ENHANCED DESIGN FLOW

We elaborate on the latest additions to MCF shown in Figure 8.1, which enhances the correctness of the IP composition process enabled through our framework. Note that the IP reflector and restricter ingredients have not altered in the newer design flow.

Modeling Framework

In addition to the CCL for specifying the architectural template, we allow behavioral descriptions through the Property Specification Language (PSL). These descriptions are provided along with the template at the component level as properties expressing its temporal characteristics. The behavioral properties of a component are finally converted into an intermediate representation (XML-IR) using an XML schema.

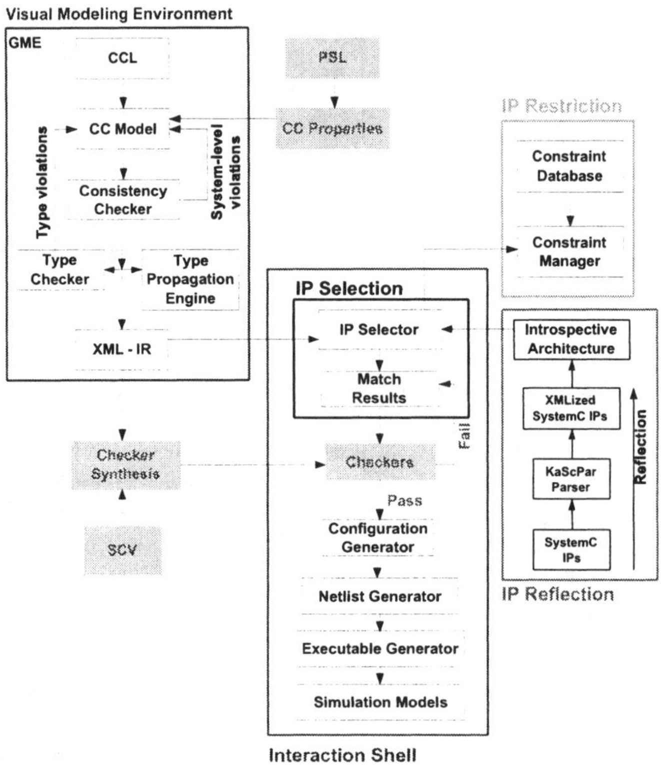

Figure 8.1 Enhanced design flow diagram.

Interaction Console

In the previous flow, the IP composition process primarily relied on structural characteristics to select matching IPs, but the newer flow has additions that remove IPs that do not behaviorally match the template. This is enabled through the properties' descriptions in the architectural template, which are used to generate checkers. These are run on the results obtained from the structural selection and the IPs that do not satisfy the properties are deemed a mismatch and removed from the result set. Note, we take the simulation-based approach to validating the IPs, but if an IP satisfies all the properties, then it can only be said to a possible match. This approach does not guarantee an exact match.

8.2 ENHANCED MODELING FRAMEWORK

The modeling framework facilitates the creation of an architectural specification of the system design. The specification process allows the description of a system template consisting of components and their interactions, where the components are placeholders for actual IP blocks. The description of a component in the template has a structural and behavioral aspect to it. The structural aspect is constrained by the abstraction level of the specification. For an RTL component, the structural description is provided in terms of input-output ports, parameters, and hierarchical embeddings. The structure of a TL component is specified using transaction read-write functions, interfaces, and channels. The behavioral description of a component is provided through rules that express a known relation between the ports or function calls. These rules specify: (1) When does a behavior occur, (2) what is the cause and effect of a behavior, and (3) how do behaviors interact? An illustration of a component description is shown in Figure 8.2.

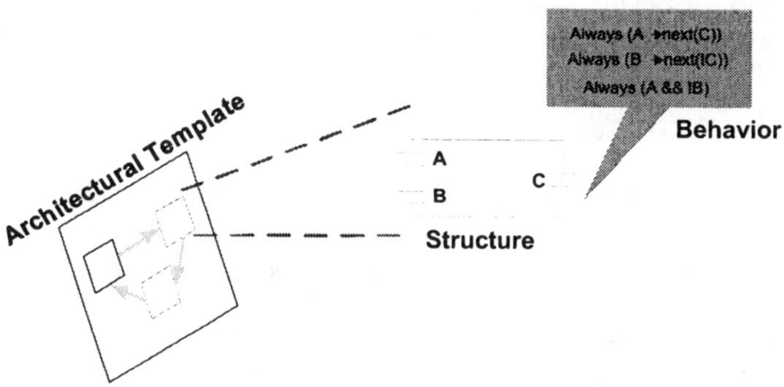

Figure 8.2 Architectural template.

The structural description is enabled through the component composition language. The behavioral description is expressed as properties using the property specification language (PSL) [3]. These properties are called the component composition properties (CCPs).

A property constrains the behaviors of a component. PSL is used to specify properties of the design and also indicate how a verification tool should use these properties in ensuring correctness of the design. PSL enables the designer to specify assertions and assumptions on the design. An assertion states that the design in question is expected to behave as described by the property. An assumption is a constraint on the environment, which is an expectation of the component regarding its neighbors.

In the MCF modeling framework, the assertion language of PSL is employed in expressing properties that capture temporal relation between: (1) inputs and outputs for an RTL component, and (2) reads and writes for a TL component. We specify the behavioral properties using a limited subset of the PSL constructs, mostly linear temporal logic and do not use its full capability. These properties are utilized by the downstream tools for: (1) synthesizing complex compositional properties, (2) synthesizing conformance tests for validation, and (3) generating traces.

Listing 8.1 Properties for the controller component of the FIR from Figure 7.7(b).

```

// Properties from the controller block
// Property 1
assert (!CLEAR -> next(OUT == 4))
// Property 2
assert always (!CLEAR && (OUT == 4) -> next(OUT == 1))
// Property 3
assert always (!CLEAR && !IN_VALID && (OUT == 1) -> next(OUT == 1))
// Property 4
assert always (!CLEAR && (OUT == 1) && IN_VALID -> next(out == 2))
// Property 5
assert always (!CLEAR && (OUT == 2) -> next(OUT == 3))
// Property 6
assert always (!CLEAR && (OUT == 3) -> next(OUT == 4))
```

Listing 8.2 Properties for the convolutor component of the FIR from Figure 7.7(b).

```

// Properties from the convolutor block
// Property 7
assert always (CLEAR -> ((VALUE == 0) && !OUT_READY))
// Property 8
assert always (!CLEAR && (OUT == 1) -> (!OUT_READY))
// Property 9
assert always (!CLEAR && (OUT == 2) -> (!OUT_READY))
// Property 10
assert always (!CLEAR && (OUT == 3) -> (!OUT_READY))
// Property 11
assert always (!CLEAR && (OUT == 4) -> (OUT_READY))
```

8.2.1 Component Properties

The PSL properties are provided through the CCL, where every virtual component in the architectural template has an attribute to capture the CCPs. Consider the architectural template of an FIR core shown in Figure 7.7(a). The **fir_computation** constitute two blocks, namely **controller** and **convolutor**, which are shown in Figure 7.7(b). The behavioral properties specified by the designer on the FIR constituents are shown in Listing 8.1 and Listing 8.2.

8.2.2 Composition Properties

The properties for individual components are provided by the designer. The properties regarding the component composition are automatically synthesized. Consider the composition of properties $prop_i$ and $prop_j$ with the assumption that these express relations between the inputs and outputs of two components. The steps involved in synthesizing the compositional properties are shown here:

1. The outputs of $prop_i$ and inputs of $prop_j$ become internal variables as a result of the composition.
2. Replace an internal variable in $prop_j$ by $prop_i$, if $prop_i$ asserts the current state of that variable.
3. Repeat procedure with a different property until all internal variables are replaced.

Illustration of the property synthesis is shown in Listing 8.3.

Listing 8.3 Illustration of compositional property synthesis.

```

// Composition Properties Automatically Synthesized
// Property 12: Composition of Property 1 and Property 2
assert (!CLEAR && next[1](!CLEAR) -> next (OUT == 1))

// Property 13: Composition of Property 7 and Property 8
assert (!CLEAR && next[1](!CLEAR) && next[2](!CLEAR) -> (!OUT_READY))

// Property 14: Composition of Property 4 and Property 9
assert always (!CLEAR && inp_valid -> next (OUT == 2))

// Property 15: Composition of Property 5 and Property 10
assert always (!CLEAR && IN_VALID && next[1](!CLEAR) -> next (OUT == 3))

// Property 16: Composition of Property 6 and Property 11
assert always (!CLEAR && IN_VALID && next[1](!CLEAR) && next[2](!CLEAR) -> next (OUT == 4))

// Property 17: Composition of Property 2 and Property 12
```

```
assert always (!CLEAR && IN_VALID && next[1](!CLEAR) && next[2](!CLEAR) && next[3](!CLEAR)) ->
next (OUT == 1))

// Property 18: Composition of Property 7 and Property 13
assert always (!CLEAR && IN_VALID && next[1](!CLEAR) && next[2](!CLEAR) && next[3](!CLEAR)) ->
next (!OUT_READY))
```

Note that the properties 13 and 18 specify different sets of inputs that drive the system to produce the same output. Therefore, composing component-level properties can result in many properties and the user can put a bound on the number of properties that needs to be generated. Furthermore, the composition properties can also be provided by the designer along with the component properties.

8.2.3 XML Translation

The final step performed by the modeling framework is exporting the structural and behavioral description into the intermediate representation (XML-IR) in XML. The translation of the PSL properties into XML employs a C++ parser, an XML schema (shown in the Listing 8.4), and XML data structures.

Listing 8.4 DTD for PSL specification.

```

<!ELEMENT PSLProperties (Property)+ >

<!ELEMENT Property (Assert) >
<!ATTLIST Property name CDATA #REQUIRED description CDATA #IMPLIED >

<!ELEMENT Assert (TemporalLayer) >

<!ELEMENT TemporalLayer (Always | Never | Ifthen) >

<!ELEMENT Always (BooleanLayer | Ifthen) >

<!ELEMENT Never (BooleanLayer | Ifthen) >

<!ELEMENT Ifthen (If, Then) >

<!ELEMENT If (BooleanLayer | Next) >

<!ELEMENT Then (BooleanLayer) >

<!ELEMENT BooleanLayer (Expression) >

<!ELEMENT Next EMPTY >
<!ATTLIST Next name CDATA #REQUIRED type (next | next_a) #REQUIRED exp_name CDATA #REQUIRED
               ncycle CDATA #IMPLIED fcycle CDATA #IMPLIED lcycle CDATA #IMPLIED >

<!ELEMENT Expression (Variable+, Next*, Expression*) >
<!ATTLIST Expression name CDATA #REQUIRED fvname CDATA #IMPLIED opname (AND | OR) #IMPLIED
                     svname CDATA #IMPLIED >

```

```
<!ELEMENT Variable EMPTY >
<!ATTLIST Variable id CDATA #REQUIRED name CDATA #REQUIRED field CDATA #REQUIRED >
```

A property translated using the XML schema in Listing 8.4 is shown in Listing 8.5.

Listing 8.5 Property 7 captured in the XML-IR.

```

// XML-IR snippet for FIR
<Property name ="property1" >
  <Assert>
    <TemporalLayer>
      <Always>
        <Ifthen>
          <If>
            <BooleanLayer>
              <Variable id ="v1" name ="CLEAR" field ="true"/>
              <Expression name ="e1" svname ="v1"/>
            </BooleanLayer>
          </If>
          <Then>
            <BooleanLayer>
              <Variable id ="v2" name ="VALUE" field ="0"/>
              <Variable id ="v3" name ="OUT_READY" field ="false"/>
              <Expression name ="e2" fvname ="v2" opname ="AND" svname ="v3"/>
            </BooleanLayer>
          </Then>
        </Ifthen>
      </Always>
    </TemporalLayer>
  </Assert>
</Property>
```

8.3 INTERACTION CONSOLE

The actions enabled through the original introspective console can be summarized into five different categories and executed in the order enumerated here:

- **Task 1:** Perform IP selection by matching virtual components from the architectural template with IP implementations from the library.
- **Task 2:** Identify the ideal match from the result set obtained after applying the selection techniques.
- **Task 3:** Bind virtual ports of the architectural components with concrete ports of the chosen IP implementation.
- **Task 4:** Create an executable SystemC model that is an instantiation of the architectural system template.

- **Task 5:** Generate random test-suites for verification of the executable model.

In the new flow, the improved introspective console enables most of the previous tasks; however the verification task is performed in a more directed manner as opposed to the previous black-box approach and carried out much earlier. The verification task is elevated to be the second task and achieved through checkers synthesized from the behavioral properties, which eliminates the need to perform random testing (see Task 5 in the enumeration).

Conformance Testing

The application of checkers for verification is called *conformance testing*. We employ the SystemC Verification Library (SCV) to generate checkers from the PSL properties specified at the component-level as well as from the synthesized composition properties. The primary purpose of these checkers is to verify the structural type-based selection of IPs. They ensure that the matching IPs behave as enforced by the properties specified on the virtual component. They are employed to sieve out the IPs that are behaviorally incompatible with the virtual component based on their properties. The conformance testing-based approach to IP reuse provides an enhanced guarantee of correctness when compared to the structural type-based approach.

Checker Synthesis

The two kinds of checkers generated from the specified PSL properties are the monitor process and the driver-monitor process. The monitor process observes the IP execution and checks if the desired behavior holds. The driver-monitor process provides a directed set of inputs that drives the system to a particular state and then checks if the desired behavior holds. The triggering mechanism for the monitor process is providing the clock input, whereas for the driver-monitor process it is a sequence of inputs given across multiple clock cycles. For a monitor process, the selected IP is said to be in conformance when in every cycle the outputs of the IP have the values asserted by the property. For a driver-monitor process, the selected IP is said to be in conformance when the checker drives the outputs of the IP to the values asserted by the property.

The monitor generation involves creating a test case, which is an SCV testbench that checks that the desired outputs are obtained in all the required cycles. The driver-monitor generation involves creating SCV constraints for the inputs and a test case. The test case is a TLM testbench that constrains the inputs and checks that the desired outputs are obtained in the required cycle. The outcome of the monitor/driver-monitor execution is: (1) the test failed and a countertrace is generated or (2) the test passed and the next chccker is executed.

Checker Deployment

The checkers are synthesized and deployed after performing structural type-based IP selection and obtaining the result-sets. The checkers are executed in a systematic loopy fashion to maximize the directed testing in order to enhance the guarantee of correctness of our solution. The checker deployment procedure is shown here:

Checker Deployment Algorithm

{Given an architectural template of a system design D, V is the set of virtual components, M is the set of match sets for each virtual component, such that M_v is the set of matching IPs for a virtual component v, P is the set of property sets for each virtual component, such that P_v is the set of PSL properties specified for the virtual component v, P_S is the set of synthesized composition property-sets ($P_S = \emptyset$), where $P_{S_{ij}}$ is the synthesized property-set for the composition of virtual components v_i and v_j. Finally, B is the set of behavioral match sets ($B = \emptyset$).}

Step 1 For each virtual component $v \in V$,
  Step 1.1 For each match $m \in M_v$,
    Step 1.1.1 For each property $p \in P_v$,
      Generate checker c for property p
      If ($c(m)$ = `fail`) then break;
    Step 1.1.2 insert(B_v, m);
Step 2 For each $v = v_i \bowtie v_j \in D$, where $v_i, v_j \in V$,
  Step 2.1 Synthesize S_{ij} from P_{v_i} and P_{v_j}
  Step 2.2 For each property $p_s \in S_{ij}$,
    Step 2.2.1 Generate checker c_s for p_s
    Step 2.2.2 For each match pair $b_i \bowtie b_j$, **s.t.** $b_i \in B_{v_i}$ and $b_j \in B_{v_j}$,
      If ($c_s(b_i \bowtie b_j)$ = `fail`) then remove(B_{v_i}, b_i); remove(B_{v_j}, b_j); break;
Step 3 Repeat 2 with $v = v_i \bowtie v_j \bowtie v_k \in D$, until $v = D$
Step 4 Output B;

In the first stage, the procedure synthesizes checkers for the component-level properties. These checkers are executed on the IPs selected for each virtual component. If the test fails, then that particular IP is removed from the match-set. In the next stage, the procedure synthesizes checkers for the compositional properties (⋈ represents composition in the algorithm). These checkers are executed level-by-level on the composition of the selected IPs. The procedure terminates when the level-by-level composition results in the complete design and all the component and composition properties are checked and enforced. We provide a diagrammatic example in Figure 8.3 to illustrate the deployment procedure.

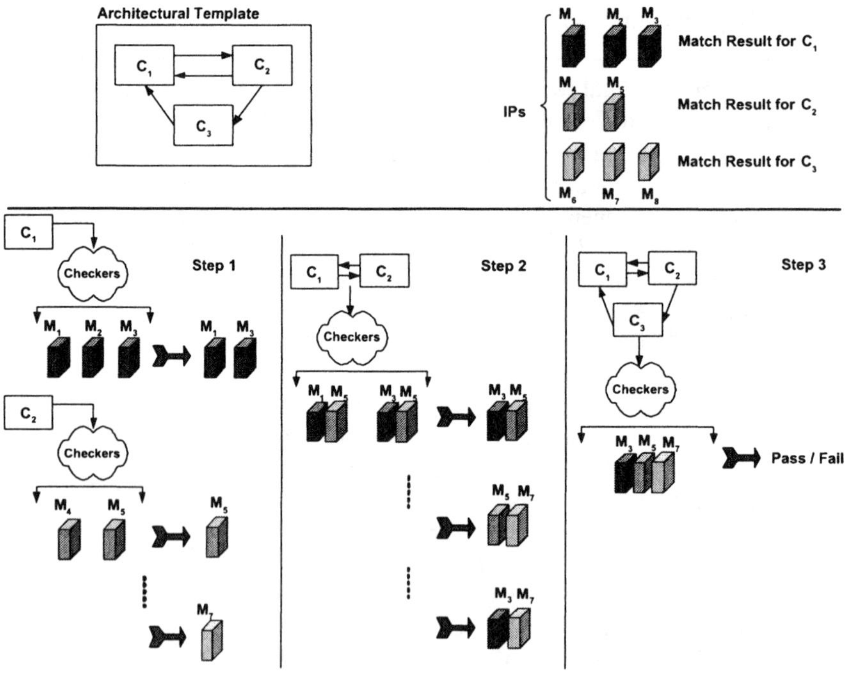

Figure 8.3 Illustration of checker deployment.

8.4 CONCLUSION

We have implemented MCF as a proposed solution to the IP reuse problem. We employ a behavioral approach (augmented to the previous structural approach). The main ingredients are: (1) visual modeling environment, (2) IP library with reflection and capability, and (3) introspective shell.

References

[1] D. A. Mathaikutty and S. K. Shukla, "SoC Design Space Exploration Through Automated IP Selection from SystemC IP Library," *Proceedings of IEEE International SoC Conference (SOCC'06)*, Austin, TX, September 2006, pp. 109–110.

[2] D. A. Mathaikutty and S. K. Shukla, "Mining Metadata for Composability of IPs from SystemC IP Library," *Proceedings of Forum on Specification and Design Languages (FDL'06)*, TU Darmstadt, September 2006.

[3] PSL/Sugar Consortium, "Property Specification Language," http://www.pslsugar.org/.

Part II

Verification Reuse

Chapter 9

A Metamodel for Microprocessors

In this chapter, we propose a visual microprocessor modeling and validation environment (MMV). MMV is built on top of a metamodeling framework which enables retargetable code generation and facilitates creating models at different levels of abstraction, without tying them to any specific programming language. This language-independent representation enables the analysis and transformation of the models uniformly into various validation collaterals. It also allows a manual refinement methodology and enforces consistency within and across abstraction levels. As an example, we model a 32-bit RISC processor at the system level, as well as architecture and microarchitecture levels to illustrate MMV's modeling capability and discuss how to extract constraints for test generation at each of these levels of abstraction.

9.1 MODELING AND VALIDATION ENVIRONMENT (MMV)

The MMV environment facilitates microprocessor modeling across various abstractions and uniformly generates code that can be targeted towards various validation platforms. The requirements for such an environment are: (1) language-independent modeling syntax and semantics, (2) a uniform way to capture the various modeling abstractions, (3) a way to express rules that enforce well-defined model construction in each abstraction level and consistency across abstractions, (4) a visual editor to facilitate the modeling based on the specified syntax and semantics and to enforce the modeling rules through checkers during design time, and (5) the capability to plug-in code generators/interpreters to allow for model analysis and translation into executables, validation collaterals, test plans, coverage tools, and so forth. The key enabling capability for MMV is *metamodeling*, which is provided through the generic modeling environment (GME). In this chapter, we demonstrate the multi-abstraction modeling framework and code generation for one specific validation target.

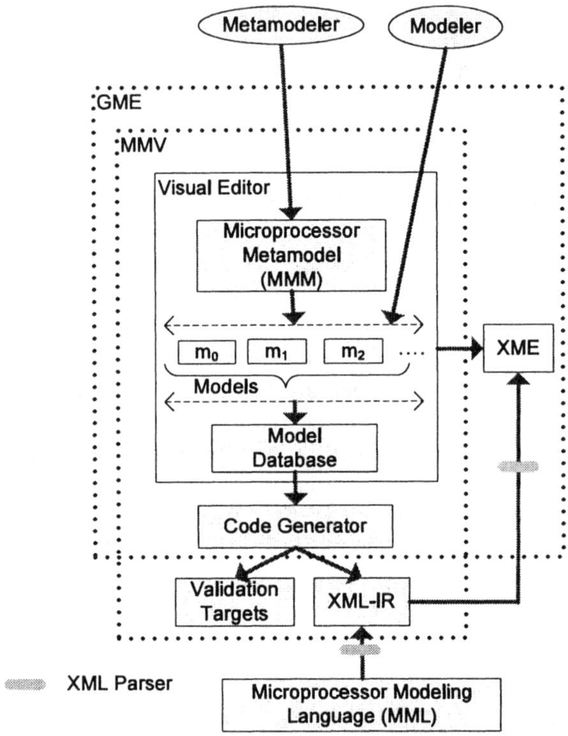

Figure 9.1 MMV environment.

Figure 9.1, shows two types of users of MMV, namely the modeler and the metamodeler. The metamodeler customizes MMV based on the requirements of the processor modeling and validation domain and also creates code generators for different validation targets. The modeler creates microprocessor models in MMV and uses the code generators to generate validation targets. When dealing with complex microprocessors, scalability of a visual modeling tool is a concern. Therefore, MMV allows two modeling modes: (1) visually based on the microprocessor metamodel (MMM) and (2) textually based on the microprocessor description language

(MDL). Having a textual input addresses the scalability issue with visual modeling tools and allows the modeler to take advantage of MMV's validation tools and checkers.

Microprocessor Metamodel (MMM)

The domain-specific syntax and some required static semantics for modeling microprocessors in different levels of abstraction is captured as a metamodel in MMV. The metamodel is built in GME using UML constructs and OCL constraints. It depicts four distinct modeling abstractions of a microprocessor and provides a unified language to enable the specification process. The usage of UML to construct the microprocessor modeling syntax and semantic provides language-independence as well as interoperability through an XML-based intermediate representation/exchange medium. It alleviates the language-specific artifacts such as pointer manipulation, strong typing in C/C++ as well as overcomes the modeling restriction of languages such as VHDL/Verilog with object-orientated features embedded within them.

The abstractions are: (1) system-level view, (2) architectural view, (3) microcode view, and (4) microarchitecture view. The modeling language provides a refinement methodology across abstraction levels where the syntactic-level transformation are well defined and enforced during refinement. Each of these abstractions corresponds to a view of the metamodel and is described using the notion of aspects in GME. Aspect partitioning provides a clean way to separate the modeling constructs within a language to allow different modeling styles. Each of these modeling styles corresponds to a distinct abstraction and the only way to cross the abstraction is to migrate to a different aspect. MMV's modeling framework provides the means to consistently merge these aspects and provide a multiabstraction microprocessor model. In the following subsection, we outline the metamodel description for some of these abstractions. The structure of the processor components is captured through the *entities* and *relations* provided in the metamodel and the behavior is embedded into *attributes* of the different entities and relations in the metamodel.

9.1.1 System-Level View (SV)

This abstraction provides a protocol-level description of various microprocessor features, which can be thought of as a software model for the processor component. The modeling framework at this abstraction provides a high-level algorithmic specification capability. All the entities describing the high-level model need not be refined to the lower abstraction, since they may not have any significance at another level and are artifacts of the SV. However most of the entities when refined to the next level, which is the architectural view, are mapped to one or more sequences of instructions or hardware architectures.

Figure 9.2 shows a snapshot of the system-level aspect of the MMM. The topmost entity of the metamodel is a **Platform**, which acts as a container for the different constituents of a processor. The constituents visible at the SV are **Core**, **Main_Memory**, and entities that describe the communication structure. The **Core** is used to describe the CPU of the processor at the system-level and the **CoreType** attribute is used to inform the framework that an existing core is being modeled. This allows the framework to enforce constraints at the current abstraction as well as across abstractions based on the characteristics of the chosen core. An example would be the Pentium Pro core's register file, which has 32-, 16-, and 8-bit registers, but some of the recent cores allow for 64-bit register. If the user is modeling at the architectural level and has set the **CoreType** to *Pentium Pro* in SV, then the framework activates a trigger that will flag a violation whenever a 64-bit register is created.

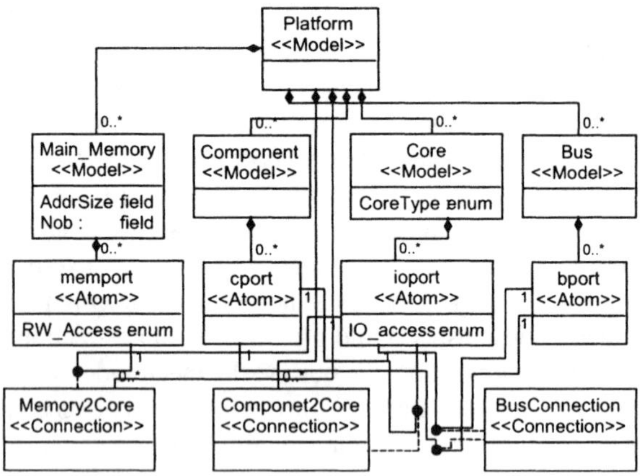

Figure 9.2 System-level view of MMM.

Another important constituent of the **Platform** is the **Main_Memory** entity, which is used to model the memory by capturing information on the address size and number of bytes/location. In SV, memory is virtual with no physical existence, therefore the attributes are only configurable at lower abstractions. This is enforced as a constraint to restrict the users from mixing the abstractions inconsistently. In

SV, we have the notion of function calls, which is modeled through ports, in order to capture the communication between the memory and the core. The core has **ioports** that could be input or output specified through **IO_access** and the memory has **memports** that could be specified as read or write ports using **RW_access**. Point-to-point communications are modeled using these ports. We also allow for bus-based communication, which is enabled through the **BUS** and **busport** entities. The entities **Memory2Core** and **BusConnection** are of type <<Connection>> in GME, which captures the communication between the memory and core as well as their communication through a bus by instantiating ports. The metamodel also has an entity that is a placeholder for some block that models external interaction with the processor and memory. The block models a functionality at the protocol-level such as arbitration and scheduling and is described using **Component**, **cport**, and **Component2Core**.

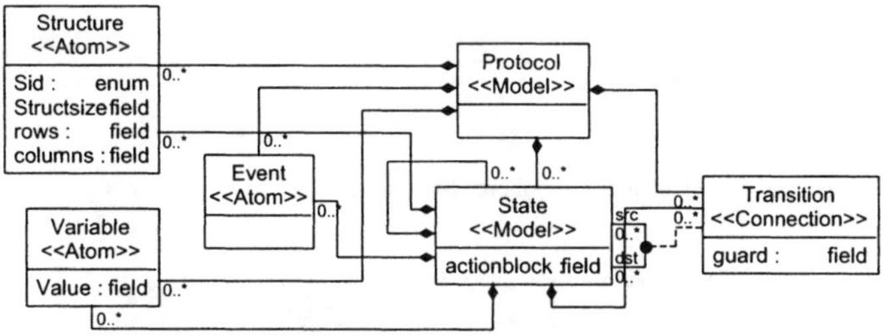

Figure 9.3 Protocol entity.

This abstraction allows the modeler to provide a protocol-level description of all entities in the SV, which is captured as a variant of hierarchical finite state machines (HFSM) in the metamodel. This was done by having the **Protocol** entity shown in Figure 9.3 contained within the **Core**, **Main_Memory**, and **Component**. The HFSM description had the addition of state variables (**Variable**) and complex structures (**Structure**) such as queue, stack, table, and so forth. It also had the notion of events, which is an alias for system-level/user-level occurrence modeled by an **Event**. They are used to mark entry/exit points for a behavior during a protocol description.

As an example, we model a possible cache behavior at the SV, specifically the MSI protocol with two cores interacting with the memory through a control hub (MCH) and bus lines shown in Figures 9.4 and 9.5. The *MCH* is modeled as a **Component** and the other elements in the toplevel are straightforward instantiations. The MSI protocol interfaces in both cores are identical and have been abstracted to a single data unit. The cores can be in one of the three possible states with respect to a cache line, namely Modified, Shared, and Invalidate as shown in Figure 9.5. The processor read and write are modeled by the events *PrRd* and *PrWr* and protocol states have access to the variables *data* and *cstate*. The *data* variable holds the cache data that is being read or modified and the *cstate* is a string capture of the protocol state, which is used in the snooping stage to set the HIT and HITM output.

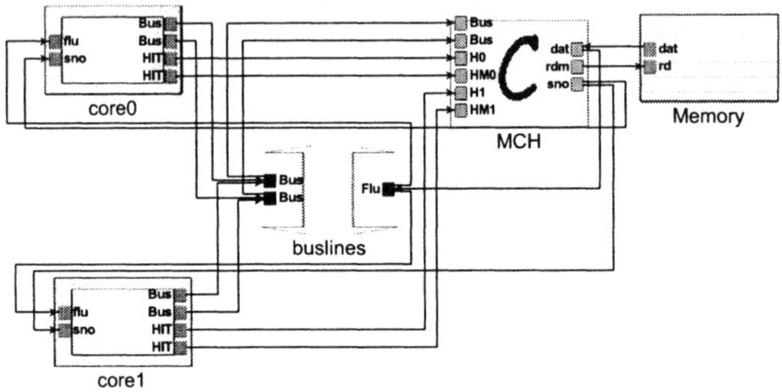

Figure 9.4 Structural SV of MSI cache coherence protocol.

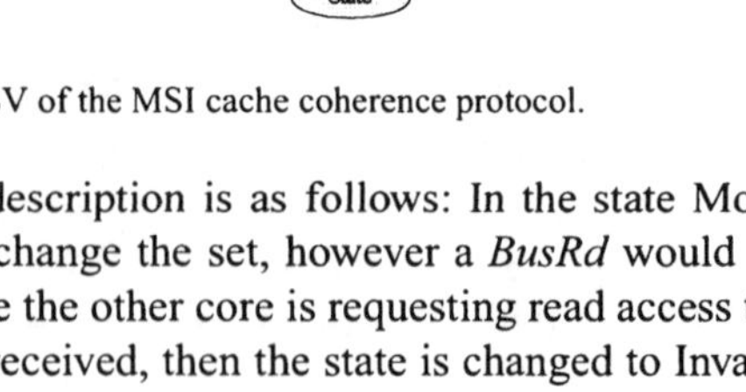

Figure 9.5 Behavioral SV of the MSI cache coherence protocol.

The protocol description is as follows: In the state Modified, a *PrRd*/*PrWr* occurrence will not change the set, however a *BusRd* would require it to move to the Shared state since the other core is requesting read access to the data in its core. But if a *BusRDX* is received, then the state is changed to Invalidate since the other core is requesting exclusive access to the data, which could be with the intent to change it. The behavior description at each state is captured as a code fragment in the **actionblock** attribute of the **State** entity and the condition on which transition is taken is captured as a **guard** attribute. The behavior at the Invalidate state is shown in Figure 9.6, where it sends out the data to the flush output line.

```
Snoop_Check:
begin
  if data.present()
  begin
   case (cstate)
   Modified :
    put(HITM) = TRUE
   Shared   :
    put(HIT) = TRUE
  end
  else
  begin
   put(HITM) = FALSE
   put(HIT) = FALSE
  end
end
```

```
Invalidate:
begin
  put(flush) = data
  cstate = Invalidate
end
```

Figure 9.6 Code fragments for behavior description.

The MCB behavior is to snoop the cores whenever it gets a read or write request. If the HIT is asserted, then the MCB fetches the data from memory and gives it to the respective core, since the core(s) is in the Shared state. However, if the HITM is asserted then MCB does not perform any fetching, since the data in memory is not valid and the processor exerting HITM is responsible for handling the r/w request. The HIT and HITM are asserted in the Snoop_Check state when the MCB is snooping the cores as shown in Figure 9.6.

9.1.2 Architecture View (AV)

Figure 9.7 shows a snapshot of the architectural aspect of the microprocessor metamodel. This provides a visual and formal specification process for the ISA as opposed to it being an architectural specification document, where each instruction is explained with English and then some C code is shown to explain its effect.

In AV, the memory abstraction is similar to SV and does not undergo a refinement but uses **mem_arch_port** to communication with the core. The core undergoes a refinement to include register definitions that are described using the **RegisterFile** entity with a *Size* attribute, which is shown in Figure 9.8. This attribute is used in the later abstractions to map the registers to the physical address.

The relations between the different entities modeled as <<connection>>s are abstracted away as seen from the illustration in Figure 9.8.

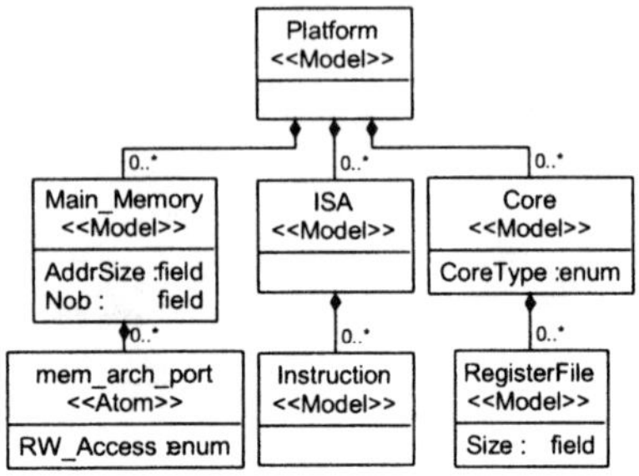

Figure 9.7 Architectural view of MMM.

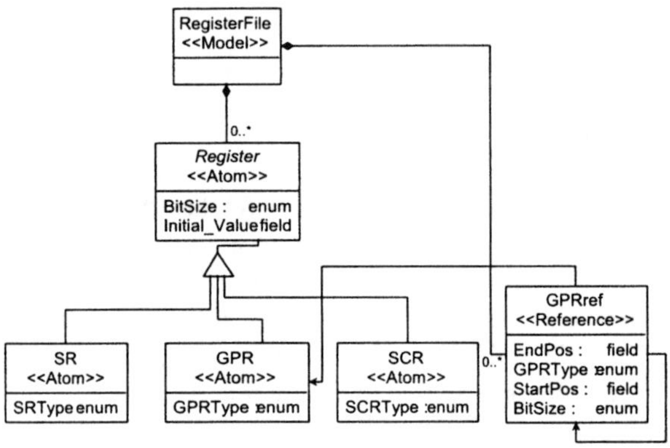

Figure 9.8 RegisterFile entity.

The **RegisterFile** allows for three type of registers, namely: (1) general-purpose register (**GPR**), (2) segment register (**SR**), and (3) statue and control register (**SCR**). Furthermore, to capture the part-whole relationship between registers we have the notion of references to registers. To model the following: a 32-bit register (say EAX) and the 16-bit part AX of EAX as well as the 8-bit parts (AL and AH) of the 16-bit part AX, we use the **GPR** and **GPRref** entities as shown in Figure 9.9.

In Figure 9.9, we instantiate a **GPR** entity to describe the EAX register and then instantiate a **GPRref** for the AX register and link it to EAX. Similarly, we

to describe any high-level behavior and are independent of programming artifacts such as pointers and datatypes. Note that only a few atomic statements are shown in Figure 9.11, but a lot more are defined in the metamodel.

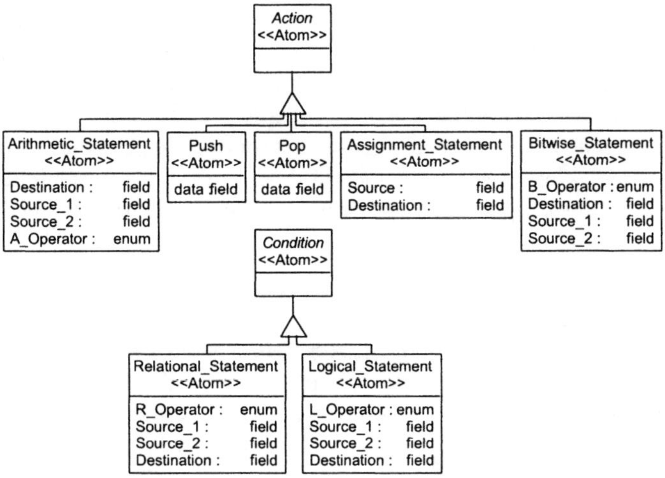

Figure 9.11 **Action** and **Condition** entities.

The statements that describe an **Action** or a **Condition** consist of one or two source operands and a destination operand. These operands are captured as attributes in the metamodel entities, namely **Source_1**, **Source_2**, and **Destination**. The **_Operator** attribute enumerates the list of possible operators for each of these statements. For example, the **A_Operator** attribute is enumerated as (+ | - | * | /). The source and destination operands of a statement entity correspond to **Identifier** entities (shown in Figure 9.10), which are instantiated and characterized before their usage within the operands of a statement. Note that these identifiers can map to registers and memory locations as well as immediate values, which are captured by the attribute **AddrMode**. If **AddrMode** of the operand is *register*, then the decode logic generated as a part of the code generation will know which register to access based on the mnemonics of the instruction. If **AddrMode** is *memory*, then the attributes **Segment**, **Offset**, and **Size** are used to denote the corresponding segment register (CS | DS | SS | ES | FS | GS in INTEL x86) and the offset within the segment and the offset size. These attributes are used to initialize the segment selector during code generation. The attribute **OpndType** of an identifier describes whether it is the $source_1/source_2/destination/source_1$ and $destination$ field of the instruction. This attribute is used to initialize the decoding logic to correctly decode the operands of the instruction based on their addressing mode given by the **AddType** attribute.

Sequentiality is not inherent in a metamodeling language, which is the case with most high-level languages such as C/C++. Therefore, additional constructs are introduced to be able to describe a sequential behavior. The construct **FunctionSeq** in Figure 9.10 is a <<Connection>>, which is used to describe a sequential specification of function descriptions, stating the order in which the functions are processed during analysis. Similar constructs exist for describing action sequences and loop sequences as well as function-action sequences, decision-action sequences, and so forth. The description of an ADD instruction using our metamodel is shown below:

Function_block: ADD
Identifier: src (**OpndType** = $Source_1$, **AddrMode** = $Register$)
Identifier: dst (**OpndType** = $Source_2$-$Destination$, **AddrMode** = $Register$)
Arithmetic_Statement: S0 := dst = src + dst

Consider the jump-if-carry instruction (JC) that positions the instruction pointer (ip) to the location specified in source (src) when the carry flag (cf) is set. The *ip* and *cf* are references to the **EIP** and **EFLAG** physical registers in the core. The *src* is an identifier that translates to a register/memory reference or an immediate value during decoding.

Function_block: JC
Identifier: src
RegisterRef: ip → EIP (**Access** = read-write, **StartPos** = 0, **EndPos** = 31)
RegisterRef: cf → EFLAG (**Access** = read, **StartPos** = 0, **EndPos** = 0)
Decision_block:
Guard:
Relational_Statement: S0 := cf == 1
Action:
Arithmetic_Statement: S1 := ip = ip + src

The **Event** entity is used to describe the state of the processor when it encounters an exception as shown in Figure 9.12. We have identified three kinds of events, namely **Fault**, **Trap**, and **Abort**. The attribute **Description** is used to provide a useful description of what raised the exception for the purpose of debugging. The attribute **Stage** is used to specify when the exception occurred, which may be a pipe stage or a user-defined point during the functional simulation. The *Priority* attribute helps in processing the events based on their importance, when multiple events occur at some state of the processor. An exception is generated when a violating condition or an illegal state is encountered during the execution of an instruction. Therefore, it is contained within an **Instruction** entity.

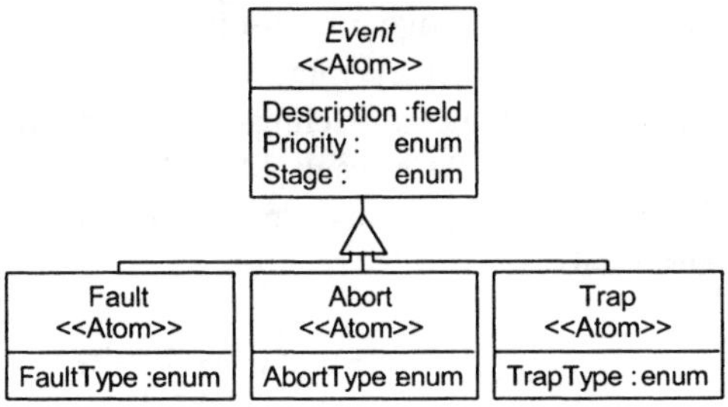

Figure 9.12 **Event** entity of the metamodel.

9.1.3 Microarchitecture View (MV)

A snapshot of the microarchitectural aspect of the MMM is shown in Figure 9.13. At this abstraction, the **Core** and **Main_Memory** undergo a refinement. The virtual memory is made concrete by defining the memory map using the **Plocation** entity providing the physical address. The core on the other hand undergoes a refinement, where the registers defined in AV are bound using a map function. In the MV, we provide a **RegisterMap** entity to map the defined registers to addresses corresponding to the **RegisterFile**. During the refinement of the Core, invalid register map functions need to be avoided, since the registers are defined in a different abstraction; this inconsistency is common. Therefore, the refinements are enforced as constraints and the modeler is required to conform to these to arrive at a consistent description of the microprocessor across the abstractions. Furthermore, a new constituent is inserted as a refinement, which is the **Pipeline** entity. A Pipeline acts as a collection of **Stages** and instruction registers (**IRegisters**).

Pipeline communicates with the memory to read and write data and address, however this communication should not be visible at SV. The **Main_Memory** should not grant read/write access to the pipeline through the **memport**, since it is meant exclusively for system-level communication with the processor core. As a result, the **Main_Memory** entity has a new set of ports, namely **mem_march_ports**, which renders this communication possible only at MV. **Pipeline** also communicates with its internal stages acting as an interface to read/write the memory and provide it to/from the different stages. The pipeline registers are modeled using the **IRegister** entity. The functionality of the stages of a pipeline vary widely, but can be described using some basic building blocks such as MUX, ALU, and adders. The MV is the most complex level in MMV and varies distinctly across processors based on whether the execution is speculative, out-of-order, and so forth. Therefore, we do not provide a framework with such elaborate modeling capability. However, one of the foremost advantages of a metamodeling-driven modeling framework such as MMV is the ease of extension. As a result, MMV's microarchitecture view can be quickly extended to incorperate memory arrays and state machines to model reorder buffers and register tables.

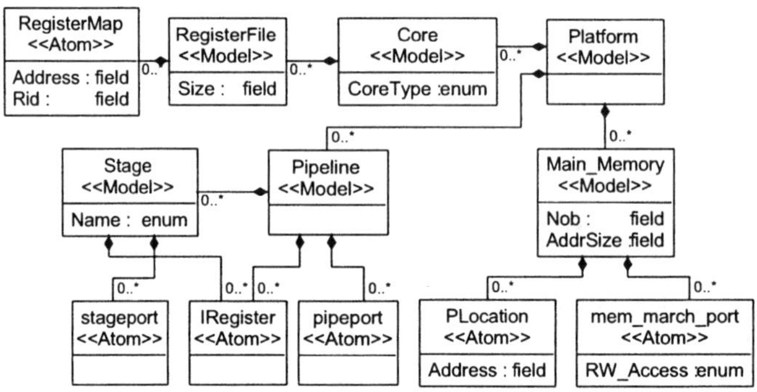

Figure 9.13 Microarchitectural view of MMM.

Another important part of the modeling framework is the constraint manager, which enforces modeling restrictions (OCL constraints) and checks for violation at model-time. Many constraints are written in MMV to enforce such static semantics and enable a correct-by-construction modeling approach. However, these are structural constraints and do not enforce any execution semantics on the model for conformance. In order to enforce certain semantics during execution, MMV allows the modeler to trace entities in the different abstractions by generating tests based on user-defined constraints. The framework is allowed to tag certain entities across abstractions and then specify equivalence constraint between them. These are enforced by generating tests that validates the equivalence. For example, a variable in the SV that is used to activate a transition to a particular state can be tagged, and during refinement the corresponding processor state that corresponds to this variable can also be tagged. The modeler can specify an equivalence constraint between the two tags that have to be enforced by the framework. Note that this capability is limited, since the SV is a high-level software abstraction and does not deal with the actual instruction execution. Therefore it is only possible to trace specific aspects across abstractions, especially if they remain the same and the components/features of the processor that use these entities have undergone refinement. In this case, the framework enforces that the common entities are used in such a way that an equivalence can be established. The notion of equivalence is model-specific and needs to be provided by the modeler.

Target Generation

During modeling, the metamodel entities are instantiated, characterized, and connected to describe the microprocessor model. These entity instances and connections are collected and populated into a model database. GME provides appropriate API calls to interface with the database, which facilitates easy access to model elements and a way to perform analysis. In MMV, model interpretation is carried out and the result is a translation of the model into some target that allows microprocessor validation. Some of the common targets are functional simulators, test generator, and coverage tools. During translation the model is analyzed in its current abstraction as well as across abstractions and the elements are extracted into homogeneous sets. These sets are treated uniformly during target-specific code generation. Therefore, our translation has two stages: (1) extraction process and (2) target-specific code generation as shown in Figure 9.1. We illustrate a part of the algorithmic description of the extraction process below. The target independent code generator is called for the second stage of the translation. In Section 9.2, we discuss how the instruction models at the AV are converted into executable C++ descriptions (***ALGS***) by the code generator. In Section 9.3, we discuss how the models are converted into constraint satisfaction problems for test generation by the code generator.

Extraction Process

{Given a model M, ISET = FNSET = IFELSESET = GSET = ASET = STSET , . . . = $\emptyset$}
Step 1 Extract entities of type **Core** and populate CORESET
For each $c \in$ CORESET,
SV : Extract entities of type **ioport** and populate IOPORTSET
For each $p \in$ IOPORTSET, call TICG(p);
AV : Extract entities of type **Register** and populate REGSET and initialize *regfile* with the **RegisterFile**, instance

For each $r \in$ REGSET, call TICG(r);
call TICG(*regfile*);
MV : Extract entities of type **RegisterMap** and populate REGMAPSET
For each instance $r_{mp} \in$ REGMAPSET, call TICG(r_{mp});
call TICG(c);

Step 2 Initialize *mem* with the **Main_Memory** instance
SV : Extract entities of type **memport** and populate MEMSET
For each $p_m \in$ MEMSET, call TICG(p_m);
AV : Extract entities of type **mem_arch_port** and populate MEMASET
For each $p_{ma} \in$ MEMASET, call TICG(p_{ma});
MV : Extract entities of type **mem_march_port** and populate MEMMSET
For each $p_{mm} \in$ MEMMSET, call TICG(p_{mm});
Extract entities of type **PLocation** and populate PLOCSET
For each $l \in$ PLOCSET, call TICG(l);
call TICG(mem);

Step 3 Initialize *pipeline* with the **Pipeline** instance,
MV : Extract entities of type **pipeport** and populate PPORTSET
For each $p_p \in$ PPORTSET, call TICG(p_p);
Extract entities of type **Stage** and populate STAGESET
For each $s \in$ STAGESET,
Extract entities of MUX, ALU, INC and populate the MUXSET, ALUSET, and INCSET
For each ($m \in$ MUXSET, $a \in$ ALUSET, $i \in$ INCSET), call TICG(m),
Extract entities of type **IRegister** and populate IREGSET
For each $r_i \in$ IREGSET, call TICG(r_i);
Initialize *isa* with the **ISA** instance,
AV : Extract entities of type **Instruction** and populate ISET
For each *inst* $\in$ ISET,
Extract entities of type **Function_Block** and populate FNSET
For each *fn* $\in$ FNSET,
Extract entities of type **Decision_Block** and populate IFELSESET
For each *dec* $\in$ DECSET,
Extract entities of type **Guard** and populate GSET
For each $g \in$ GSET,
Extract statements and populate STSET
Extract entities of type **Action** and populate ASET
For each *act* $\in$ ASET,
Extract statements and populate STSET
For each *st* $\in$ STSET, call TICG(*st*);

```
        ...
        call TICG(dec);
      call TICG(fn);
    call TICG(inst);
```

Note that the extraction process blindly follows the containment hierarchy and aspect partitioning in MMM and populates the respective sets. These sets are not shared between abstractions and this lets the code generator be oblivious to the abstraction level. The TICG is a templated function that is an integral part of the code generation and is very target-specific. For test generation based on constraint satisfaction problems (CSPs), the code generator internally constructs a program flow graph and perform single static assignment (SSA) analysis on it. The outcome of this analysis is used to print out CSPs. The details are provided in the following section.

The code generation also translates the microprocessor models into an intermediate XML representation (XML-IR), which has a well defined schema. Listing 9.3 shows the XML-IR generated for an ADD instruction. The XML-IR functions as a middle ground for the visual modeling framework and the textual interface provided through MDL. MDL has a one-to-one correspondence with XML-IR and a simple Perl parser provides the conversion. Similarly, the XML-IR is converted using a nontrivial XML parser into XME that MMV can visualize.

Listing 9.3 XML-IR for add instruction.

```
<Instruction name = "ADD" Textual = "No">
  <Function_block name = "add" >
     <Identifier name = "src" OpndType = "Source1" AddrMode = "register" />
     <Identifier name = "dst" OpndType = "Source2_Destination" AddrMode = "register" />
     <Arithmetic_Statement name = "S0" Source1 = "src"  Source2 = "dst" Destination = "dst"
      AOperator = "+" />
  </Function_block>
</Instruction>
```

XML-IR can be easily converted into another format called XME, which is the input language for GME using XML parsers. This provides an alternate input flow for our tool, which allows the modeler to textually describe the processor using XML. This textual model is converted into XME that MMV can visualize for the modeler. However modeling in XML is tedious and not a common practice. Therefore, we provide a textual microprocessor description language (MDL). It allows multiabstraction microprocessor modeling by providing syntactic constructs for the XML schema.

Advantages of MMV

The primary advantage of the tool is its metamodeling capability, which from the metamodeler's perspective makes MMV highly customizable and an easily extendible environment. They can easily add newer abstractions such as the RTL or customize an abstraction by introducing constructs specific to their processor domain such as the need to have network-on-chip capability at the SV. From the modeler's perspective the tool allows a correct-by-construction approach through their modeling-time constraint checks as well as provides a neat and compact way of consistently visualizing their model across various abstractions. An XML-based representation of the models allows an alternate textual input capability for the tool. This caters to the need of the modelers who are not comfortable with visual modeling, but want to take advantage of our refinement methodology, consistency checkers, and validation targets. The next most important capability of the tool is the multitarget code generation. GME captures the model in a database and provides APIs to efficiently access different model elements and develop plug-ins that perform target-specific translation. It again promotes extendibility by allowing the metamodeler to add newer application-specific targets.

9.2 SIMULATOR GENERATION

The simulator generation in MMV is performed through a code generator that translates the various instruction models into *ALGS*. These are executable descriptions of the instructions with the operations mapped out and the register and memory accesses resolved. The objective at this point is not to generate a full-fledged architectural model but to extend an existing architectural simulator for futuristic architectural requirements. Some of such requirements are increasing instruction or cache sizes, adding new register states, and so on. These extensions to an existing architectural specification can not be propagated onto a functional model without reworking a lot of the development. This difficulty arises because most available architectural models were developed by architects and not software engineers. Therefore, the development was specifically to fulfill the requirements of the specification without much thought on how the model would adapt to the newer requirements. For example, the IA-64 instructions are capable of handling up to 64-bit operands. If the next generation processor requires 128-bit instructions, then the developer facilitates this feature by modifying those instructions of the architectural model. Note that many times the modification touches most if not all instructions and therefore is a tedious task. Our objective is to ease this task by demonstrating how to generate extensible architectural models. We employ the metamodeling capability to describe generic architectural concepts and capture information on the target and architectural specifics to enable the construction of an executable model of the ISA.

9.2.1 Metamodel Additions for Simulator Generation

The AV metamodel also has certain entities that are used to capture architectural metainformation that are exploited during simulator generation. The entity **Metadata** is the key ingredient of our metamodeling-driven extensible simulator generation approach. This entity captures the architectural-specifics of a processor as well as the target-specifics of the underlying programming language. Once the architectural details are provided through the **Metadata** entity, our generic metamodel is configured and customized to a specific processor-metamodel. Once the programming language artifacts are provided through the **Metadata** entity, the code generator is initialized to generate an executable in that language. Every architectural construct with an enumerated attribute can be subjected to specialization through the **Metadata** entity. Examples of some architectural details captured through the **Metadata** entity are number of registers and their sizes, instruction sizes, fault types, and event priorities, which vary across processors. On the other, programming artifacts are specifics of the target language used to create the executable model. Examples of some artifacts captured through the **Metadata** entity are typing requirements, data structures, machine size, character encodings, and endianness, which are heavily dependent on the underlying programming language and its representation. Some of these also vary from processor to processor. For example, x86 processors from Intel use the little-endian format, whereas the PowerPC from IBM adopts the big-endian format. We illustrate in Figure 9.14 the configurable code generation with respect to one specific architectural detail, namely the register sizes (**RegSizes**) and one specific programming artifact, namely the size and type of the data unit (**DataUnit**) for C++ code generation.

The attribute **Sizes** of **RegSizes** is used to capture the byte sizes of the possible registers of an appropriate architecture. For example, consider the x86 register set shown below, the entity **RegSizes** has its attribute **Sizes** specified as 1;2;4.

```
32 bits : EAX, EBX, ECX, EDX
16 bits : AX, BX, CX, DX
8 bits : AH, AL, BH, BL, CH, CL, DH, DL
```

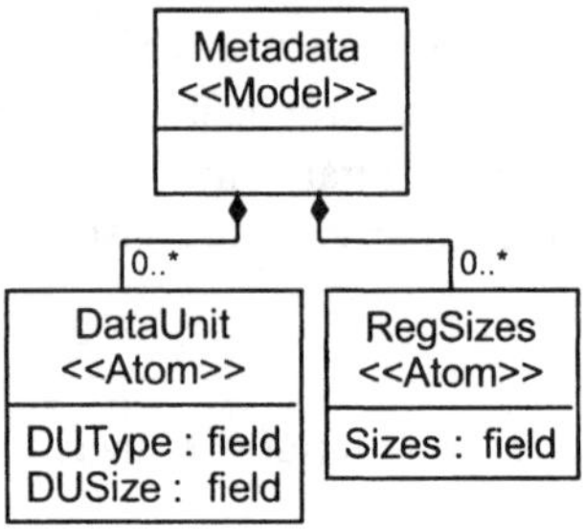

Figure 9.14 Part of the **Metadata** entity.

The entity **DataUnit** captures the data representation format that must be employed for defining the processor states. The attributes **DUType** and **DUSize** capture the bitsize and C++ datatype for the required data unit, which allows for uniform representation of all registers and memories as well as intermediate states of the processor. For example, consider the data unit definition shown below, then every state of the processor is defined using **DUType** = **DU32**, which is internally a 32-bit unsigned integer. The **DUSize** attribute for this example is 32. This attribute is employed in the manipulation of processor states with bitsizes larger/smaller than **DUSize**.

```
typedef unsigned int DU32;
```

The entity **Main_Memory** is used to model the main memory of the processor. The level of details of the memory schematic varies widely across processors. Some of the aspects that are captured through this entity are the memory hierarchy, types of memory, address ranges, word size, read-write latencies, cache replacement, and write back policies. The memory can be thought of being translated to a very large 1-D array in C++ during code generation.

9.2.2 ALGS Generation

The task is to code generate a C++ implementation of the architectural model of the processor that can be plugged into Intel's (golden) Instruction Set Simulator (ISS). The ISS is the reference implementation of the ISA, which started as a single-core, single-threaded model and was adopted to model multithreaded and multicore processors. This complexity results in modeling challenges, especially with the pace at which Intel is coming up with new architectural extensions, microarchitectures, and proliferations. Therefore, extending the ISS to incorporate new processor requirements is difficult, especially if the new requirements modify a large set of instructions. These modifications are highly error-prone since they can lead to inconsistencies. Furthermore, the process is very time-consuming and tedious, requiring several man-months for the integration. Our objective is to automate some of these tasks and alleviate the above problems associated with simulator extension.

The ISS is primarily employed for validation by different teams across Intel, therefore it has a well-defined interface. The interface provides APIs to simulate the architectural model in various processing modes, to cosimulate with other tools, as well as to access and modify the states of the processor at any given point of the simulation. These APIs provide read/write functions to either dump out the state of the processor or drive the processor to a state of interest. We employ the APIs and the metamodeling capability of our framework to illustrate an extensible code generation methodology. We code generate an ALG that is a C++ implementation of an instruction using the APIs of the ISS to read and write from the current register and memory states of the processor as per the behavior. In our future work, we will code generate the states of the processor and the APIs as well.

It is a two-part process, where first we code generate from the metamodel and then from the processor model. We translate all the atomic operators in the metamodel to C++, by generating functions that perform the operation. References to these operators in an instruction model are replaced by their appropriate function (generated from the metamodel) calls during the ALG generation.

CG from Metamodel

In this stage, the code generation is performed from the generic constructs in the metamodel, which is independent of how the modeler utilizes these constructs. The code generation employs the **Metadata** entity to obtain the specifics of the platform and the architecture. We illustrate the code generation of the **Arithmetic_Statement** operator in Listing 9.4. For this **RegSizes** is assumed to have value **1;2;4** and **DataUnit** has value **DU32** and **32**.

The **add_opr** function has five arguments, which get their datatypes from the **DataUnit** entity. The function arguments *Source_1*, *Source_2*, and *Destination* correspond to the attributes of the **Arithmetic_Statement** entity. The **Size** argument maps to the *Sizes* from the **RegSizes** attribute. Note that with **RegSizes** as **1;2;4**, it allows for 8-, 16-, and 32-bit operands for the instructions. However for the next generation processors such as IA64, we need 64-bit extensions, which requires many if not all the instructions to support 64-bit operation. Such an extension to IA32 is difficult and extremely tedious and error-prone. We illustrate how our code generation approach facilitates an easy extension. The modeler specifies **RegSizes** as **1;2;4;8** in order to take this new requirement into consideration. The addition of **8** to the **RegSizes** generates the addition clause shown in Listing 9.5 to the *case statement* shown in Listing 9.4.

Listing 9.4 **Arithmetic_Statement** with **A_Operator** = +.

```
void add_opr(DU32 *Source_1 , DU32 *Source_2 , DU32 *Destination , DU32 Size)
{
  switch (Size)
  {
    case 1: Destination[0] = (Source_1[0] + Source_2[0]) & 0xff; break;
    case 2: Destination[0] = (Source_1[0] + Source_2[0]) & 0xffff; break;
    case 4: Destination[0] = Source_1[0] + Source_2[0]; break;
    default: ErrorNotice(``Size Not Supported");
  }
}
```

Listing 9.5 64-bit version of **Arithmetic_Statement**.

```
case 8:
 DU32 Temp[2];
 DU32 Carry;
 Temp[0] = Source_1[0] + Source_2[0];
 Destination[0] = temp[0];
 Carry = compute_carry(Source_1 , Source_2 , Temp, 4);
 Temp[1] = Source_1[1] + Source_2[1] + Carry;
 Destination[1] = Temp[1];
break;
```

The underlying data format is 32 bits; therefore, two 32-bit units are employed to support the 64-bit extension. As a result during the addition operation of individual units, the carry must be taken into consideration. We employ the **compute_carry** function for this purpose. The behavior for the carry updation is also provided through our modeling framework as shown below. Note that the **carry_fn** description is not provided as a part of any instruction, but as an operator which is code generated from the model rather than the metamodel.

Function_block: carry_fn
Identifier: src1, src2, dst, and size
RegisterRef: cf → EFLAG (**Access** = write, **StartPos** = 0, **EndPos** = 0)
BitWise_Statement: S1 := src1 & src2
BitWise_Statement: S2 := src1 & !dst
BitWise_Statement: S3 := src2 & !dst
BitWise_Statement: S4 := S1 | S2
BitWise_Statement: S5 := S4 | S3
Arithmetic_Statement: S6 := size - 1
Unary_Statement: S7 := 2 ∧ S6
Arithmetic_Statement: S8 := S5 & S6
Shift_Statement: S9 := S8 >> S6
Assignment_Statement: cf = S9

The data formats are very specific to the target onto which the reference model is implemented. Consider that the data unit changed from **DU32** to **DU16**, which is defined below:

```
typedef signed short DU16;
```

This implies that the underlying data format is **DU16** (**DUType** = **DU16**), which is **16** bits (**DUSize** = **16**) in size and is of type signed short. The ISS customization to work with the new set of target requirements is not easy and involves a lot of recoding. We easily achieve this customization by retargeting our code generator through the **Metadata** entity. In Listing 9.6, we illustrate the automated customization of the **add_opr** function (Listing 9.4) that is provided by the code generator.

CG from Model

This stage begins with the secondary operators and then proceeds to the instructions. Secondary operators are descriptions that are written using the metamodel and facilitate certain common architectural tasks, such as updating the status registers, and so forth, which are performed by a large set of instructions in the same manner. We factor out these common tasks and specify them separately as the next set operators available for architectural modeling. After generating the operators, the tool generates an ALG for each instruction provided in the ISA model. An instruction described using the **Instruction** entity is a collection of entities that are collected and translated during the traversal of the various sequencing constructs

Listing 9.6 Customization of the **add_opr** function.

```
void add_opr(DU16 *Source_1, DU16 *Source_2, DU16 *Destination, DU32 Size)
{
  switch (Size)
  {
    case 1: Destination[0] = (Source_1[0] + Source_2[0]) & 0xff; break;
    case 2: Destination[0] = Source_1[0] + Source_2[0]; break;

    case 4:
     DU16 Temp[2];
     DU16 Carry;
     Temp[0] = Source_1[0] + Source_2[0];
     Destination[0] = Temp[0];
     Carry = compute_carry(Source_1, Source_2, Temp, 2);
     Temp[1] = Source_1[1] + Source_2[1] + Carry;
     Destination[1] = Temp[1];
    break;

    default: ErrorNotice(``Size Not Supported");
  }
}
```

employed in the description. We provide a procedural description of the ALG generation shown below:

The functions ($TICG_{*}(e)$) translate the entity into its equivalent C++ constructs and ISS APIs. The translation into C++ is straightforward, where a **Function_block** is translated to a function definition by $TICG_{FN}$, a **Decision_block** into an if-else condition by $TICG_{DN}$, and finally the **Loop_block** into a for-loop by $TICG_{LP}$.

The ISS has an instruction class, which is used to record the operands, addressing modes, opcode, mnemonics, and prefix, of an instruction. The tool generates the appropriate function calls to correctly populate an object of this class and employs it in describing how the current instruction executes. Furthermore, the tool utilizes ISS APIs to read as well as update the states of the processor during the execution.

Algorithm 1: ALG Generation

{Given the instruction model and the populated sets ISET, FNSET, ASET, EVSET, RREGSET, DECSET, GSET, LPSET, and so forth, from the extraction procedure}

For each *inst* ∈ **ISET**,

Step 1 For each $fn \in$ **FNSET**,
$TICG_{FN}(fn)$;
Step 1.1 For each $id \in$ **IDSET**,
$TICG_{ID}(id)$;
Step 1.2 For each $act \in$ **ASET**,
$TICG_{AC}(act)$;
Step 1.3 For each $ev \in$ **EVSET**,
$TICG_{EV}(ev)$;
Step 1.4 For each $rreg \in$ **RREGSET**,
$TICG_{RR}(rreg)$;
Step 1.5 For each $dec \in$ **DECSET**,
$TICG_{DEC}(dec)$;
Step 1.5.1 For each $gd \in$ **GSET**,
$TICG_{GD}(gd)$;
Step 1.5.1.1 For each $cd \in gd$,
$TICG_{CD}(cd)$;
...
Step 1.6 For each $loop \in$ **LPSET**,
$TICG_{LP}(loop)$;
Step 1.6.1 For each $e \in loop$,
...
...

The $TICG_{ID}$ translates an **Identifier** entity into **FETCH** and **WRITE_BACK** function calls that populate the instruction object and update the state of the processor after execution. In Listing 9.7, we illustrate the ALG generated for an ADD instruction. The FETCH function extracts the source operands from the instruction and initializes the object. Note that this extraction is dependent on the **OpndType** and **AddrMode** attributes as well as instruction-specific information such as opcode, prefix, and mnemonic that is hard-coded in the ISS for the currently known instructions. For the new instruction in the future, this information will be captured through the modeling framework.

The $TICG_{AC}$ and $TICG_{CD}$ functions replace an **Action** and **Condition** entity with calls to the appropriate function generated from the metamodel and further correctly parameterize the call with the attributes of the entity. Note that in Listing 9.3, the ADD instruction has an action (**Arithmetic_Statement**) that is replaced by a call to the **add_opr** function shown in Listing 9.4. The T_{RR} translates the accesses to the registers described by **RegisterRef** entities in the instruction into **Get_Reg_ / Load_Reg_ / Store_Reg_** function calls. Note that after the execution of

Listing 9.7 Algorithm for ADD instruction.

```
void ADD_ALG( instruction *I);
{
   FETCH(I, *I.opnd1, TRUE, *I.Source_1);
   FETCH(I, *I.opnd2, FALSE, *I.Source_2);

   //Employs operators from the metamodel
   add_opr(*I.Source_1, *I.Source_2, *I.Destination, *I.Size);

   //Employs seccondary operators
   Store_Reg_EFLAG_Overflow(compute_overflow(*I.Source_1, *I.Source_2, *I.Destination, *I.Size));
   Store_Reg_EFLAG_Carry(compute_carry(*I.Source_1, *I.Source_2, *I.Destination, *I.Size));
   Store_Reg_EFLAG_AuxCarry(compute_auxcarry(*I.Source_1, *I.Source_2, *I.Destination, *I.Size));
   Store_Reg_EFLAG_Sign(compute_sign(*I.Destination, *I.Size));
   Store_Reg_EFLAG_Zero(compute_zero(*I.Destination, *I.Size));

   WRITE_BACK(I, *I.opnd2, *I.Destination);
}
```

the ADD instruction, the status register, namely carry, overflow, sign, and so forth, are updated. This behavior is described using **RegisterRef** to these registers with **Access = write**, which is translated into a **Store_Reg_EFLAG_** function call. The assignments from registers or register lookups are translated into **Load_Reg_** and **Get_Reg_** function calls as shown in Listing 9.8 for the JC instruction.

Listing 9.8 Algorithm for JC instruction.

```
void JC_ALG( instruction *I);
{
   FETCH(I, *I.opnd1, TRUE, *I.Source_1);
   if(Get_Reg_EFLAG_Carry() == 1)
   {
      Load_Reg_EIP(*I.Destination);
      add_opr(*I.Source_1, *I.Destination, *I.Destination, *I.Size);
      WRITE_BACK(I, *I.opnd3, *I.Destination);
   }
}
```

The translation functions ($TICG_*$(e)) are applied on the entities during the traversal of the instruction model. The traversal is done along the connections starting with the initial **Function_block** entity following the **FunctionSeq** sequential construct until all the entities are translated.

In Table 9.1, we illustrate the different instruction classes and their frequencies from the x86 Benchmark that we modeled and generated ALGS through MMV.

Table 9.1
Benchmark for x86

Instruction	Frequency
Load	22
Conditional Branch	20
Compare	16
Store	12
Add	8
And	6
Sub	5
Mov reg, reg	4
Call	1
Return	1

Some of the memory variants for the Load and Store instructions and the far version of the call and return instructions require the complete memory hierarchy and the call descriptors, privilege levels, and processing modes of the processor to be modeled for the purpose of code generation.

9.3 TEST GENERATION

MMV facilitates automatic test generation as one of its validation targets. The TICG converts the description into a constraint satisfaction problem (CSP), which is solved by the test generator using a constraint solver and formulates valid test cases. A test case is defined as a set of inputs, execution conditions, and expected outputs that are developed for a particular objective, such as to exercise a particular path in the implementation or to verify its compliance with a specific requirement.

9.3.1 CSP Formulation for Generating Random Test Cases

Many functionally equivalent CSPs can be derived from an instruction model. The TICG in our implementation follows the rules described below to convert the processor behavior into a CSP.

1. Every state element (i.e., registers or memory) that is used in the behavior is represented as a constraint variable with a finite domain [min,max]. The

constraint variables are always positive. An additional Boolean constraint variable is associated with an integer variable to act as sign bit to represent both positive and negative values where required.

2. Every module in the processor model has a set of input and output constraint variables. The input constraint variables represent state elements that are *used* in the module and output constraint variables represent state elements that are *modified* by the module.
3. Every constraint variable in CSP is the target of only one assignment. If a state element is assigned a new value, a new constraint variable is generated for that state element and future references to that state element will refer to the newly generated constraint variable. This is very similar to static single assignment form (SSA) [1] used in compiler analysis.
4. The statements within the *then* block or the *else* block of if-then-else statements should be exercised only when the guard is true or false, respectively. We use a predicate variable for each statement to enforce this rule. If any state element is assigned a new value (and hence a new variable) within the *then* or *else* block, then constraints should be added such that statements following the if-then-else block will use the correct variable depending on the path taken by the instruction.

In the following subsections, we describe how the TICG applies these rules to convert the instruction descriptions of increasing complexity into CSPs.

ADD Instruction

Figure 9.15 shows the behavior of an add instruction. Let A and B represent two registers of the CPU. The description states that A and B are added and the result is stored in A. Let a_0 and b_0 be the constraint variables corresponding to the state elements A and B, respectively. If the A and B are 32-bit registers, then the domain of a_0 and b_0 is $(0, 2^{31} - 1)$. Since A gets a new value during the execution of the instruction, we define a new variable a_1 and assign the calculated value $(a_0 + b_0)$ to a_1. The equation $a_1 = a_0 + b_0$ represents the constraints of the add instruction. The variables (a_0, b_0) and (a_1, b_0) represent the input and output constraint variable set of add. (a_1 and b_0) will be the input constraint variable set for the next instruction in sequence.

Figure 9.15 shows one possible test case generated by the test generator. The constraint variables a_0, b_0, and a_1 are assigned values 5, 7, and 12, respectively.

The test generator converts this test case into a program binary that contains an `add` instruction and the registers A and B initialized to the values stored in a_0 and b_0 respectively.

ADL Description of ADD Instruction

add()
{
A=A+B;
}

Converting the Description to Constraints
Initial State Variables: a_0,b_0 Constraints: $a_1 = a_0 + b_0$; Final State Variables:a_1,b_0

A possible test case
$a_0 = 5, b_0 = 7, a_1 = 12$

Figure 9.15 ADL description, constraints, and test case for a simple `add` instruction.

Instruction with a Conditional Statement

Constraint solvers like iLog Solver allows conditional constraints of the form if-then in CSPs. We use this feature to generate constraints for instruction descriptions with conditional statements in it. Figure 9.16 shows our architectural description (ADL) of a hypothetical instruction `foo`, which has a conditional statement and an assignment statement in its description.

The constraints for this instruction is given in Figure 9.16. The constraint variables for the state elements A and B are a_0 and b_0, respectively. We also declare a Boolean constraint variable t_0 that acts as a predicate variable for the decision (i.e., condition in the *if* statement). The first two constraints relate the decision to t_0. If t_0 is set to 1, a_0 and b_0 will be assigned values such that the decision (a_0 == 0 AND b_0 == 1) is true and vice versa. Since B gets a new value within the then block, a new variable b_1 is generated for B. The next constraint relates the statement $B = 2$ to t_0. If t_0 is set to 1, then the statement $B = 2$ should be executed, which follows the semantics of the if-then statement. Hence if t_0 == 1, b_1 is set to 2, else b_1 can take any value within its domain. The state element B after the execution of `foo` is represented either by b_0 or b_1. We generate a new variable b_2 to represent the value of B after the execution of `foo` and add the final two constraints that assign b_0 or b_1 to b_2 depending on the value of t_0. This completes the constraints for `foo`.

It should be noted that the CSP can be formulated without using predicate variable t_0. t_0 is added to make the constraints more readable in complex instruction flows and for easy manipulation of constraints when generating test cases for branch coverage. The memory and performance overheads due to predicate variables were found to be negligible in our experiments.

ADL Description of foo with if statement

```
foo()
{
 if( A == 0 AND B == 1 )
 {
  B = 2;
 }
}
```

Converting the Description into Constraints

Initial State Variables: a_0,b_0
Predicate Variables: t_0
Constraints for ADL Description:
if($t_0 == 1$) ($a_0 == 0$ AND $b_0 == 1$)
if($t_0 == 0$) !($a_0 == 0$ AND $b_0 == 1$)

if($t_0 == 1$) ($b_1 = 2$)

if($t_0 == 0$) ($b_2 = b_0$)
if($t_0 == 1$) ($b_2 = b_1$)
Final State Variables: a_0,b_2

A possible test suite to get 100% branch coverage

Additional Constraint	Test Case
$t_0 = 1$	$a_0 = 0$, $b_0 = 1$, $t_0 = 1$ and $b_1 = 2$ (taken)
$t_0 = 0$	$a_0 = 2$, $b_0 = 3$, $t_0 = 0$ and $b_1 = 0$ (not taken)

Figure 9.16 ADL description of an instruction with a decision.

Instructions with Loops

Some complex instructions in modern-day processors have loops in their behavior. If the number of iterations of the loop is deterministic, the loop is first unrolled and constraints are generated for unrolled versions of the instruction behavior. An example of such an instruction is $MOVS$ in IA32 architecture, which moves data from one set of memory locations to another. For loops with nondeterministic loop counts, we use the trial-and-error method followed in [2] to generate CSP.

9.4 CASE STUDY: MODELING VESPA IN MMV

We illustrate the modeling of Vespa [3], a 32-bit RISC processor using MMV at the system, architecture, and microarchitecture abstractions.

9.4.1 System-Level Model

Vespa System Model

We model the interaction of the real-time software scheduler, processor, and timer of Vespa at the system level, which is shown in Figure 9.17. The state machine description of each component is shown in Figure 9.18.

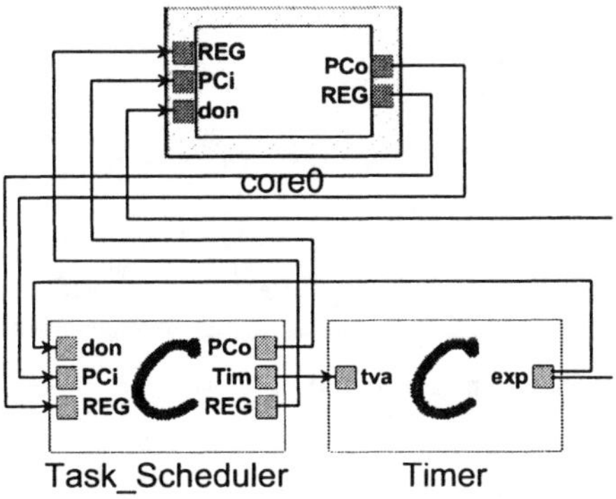

Figure 9.17 System-level model of Vespa.

Timer: It is modeled as a down counter with three states, namely INIT, DECREMENT, and EXPIRED. The timer has an internal variable *tcnt* that holds the counter value. It is modeled as a **Variable**, which is initialized by an input in the INIT state. The timer has three inputs: (1) timer's initial value *tval*, (2) *load* flag, and (3) clock. It also has an output called *expiry* that is used to say when the timer has reached zero. These are modeled as **ioports**. The timer's internal *tcnt* is initialized with *tval* when the *loadtimer* input is activated. Then the timer moves into DECREMENT, where in the subsequent clock cycles the *tcnt* becomes *tcnt* - 1. When *tcnt* becomes zero, the timer transitions to EXPIRED and the output *expiry* is set. In this state if the $load$ is activated, the timer will deassert $expiry$ and jump to INIT.

Processor: It is modeled as an FSM with three states: INIT, EXECUTE, and SUSPEND. When the processor receives the $done$ input, it goes to the SUSPEND state and sends the status of the executing task to its output ports pPC_o and $pREG_o$. On receiving the $load$ input, the processor will transition to the INIT state and initialize PC and REG with the values in the input ports PC_i and REG_i and then transitions to the EXECUTE state to begin executing the new task.

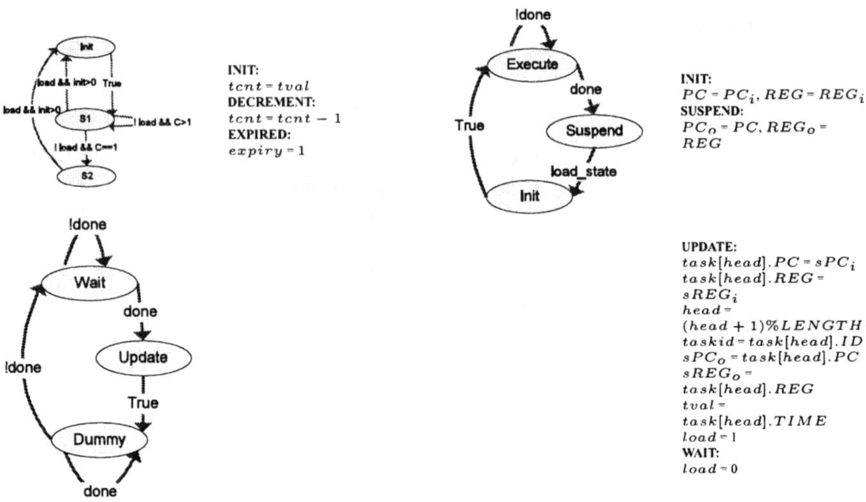

Figure 9.18 Behavior of the timer, processor, and scheduler.

Scheduler: The scheduler controls which task will execute in the processor core at any given time. We model a simple static scheduling algorithm in the scheduler [4]. All the tasks have a unique ID and are assumed to be periodic with the time taken to execute any task being constant and deterministic. The scheduler maintains the details of all the tasks in a task queue *task* modeled as a **Structure**. When the scheduler receives the timer done input (*expiry*), it wakes up and saves the state of the suspended task in its task queue. It increments the head pointer to point to the next task in the queue and sends the PC and register state of the next task to its output ports sPC_o and $sREG_o$. It sets *tval* with the period of the task and asserts the *load* signal and initializes *taskid* with the ID of the process that is going to execute.

The state machine representation at the system level hides the actual implementation of the different components. In the subsequent levels, these state machines will be manually refined to include more details. In the architecture level, the model of the processor core will be refined to include the details in the ISA and the scheduler will be implemented using the assembly language of Vespa. In the microarchitecture level, the timer will be refined as a hardware block that interacts with the pipeline of the processor using interrupts. In Sections 9.4.2 and 9.4.4 we show how the processor is refined in the architecture and microarchitecture abstractions.

Constraints for System Model

The CSP for the system model is shown in Figure 9.19. The code generator in MMV generates an *if* constraint for every transition in the FSM. The *if* constraint will guard the action performed in the state which the system will transition to. The CSP represents the behavior of the HFSM for one time step. Variables subscripted by p and n store the state before and after the time step, respectively. CSPs for multiple time steps can be generated by generating one CSP for every time step and chaining the variables. The variables representing the *next* state of one CSP will be used as the *previous* state of the next CSP.

Processor:
if($expiry_p == 1$) $pPC_{o_n} = PC_p$ && $pREG_{o_n} = REG_p$
if($load_p == 1$) $PC_n = pPC_{i_p}$ && $REG_n = pREG_{i_p}$
Timer:
if($load_p == 0$ && $tcnt_p != 0$) $tcnt_n = tcnt_p - 1$
if($tcnt_p == 0$) $expiry_n = 1$
if($load_p == 1$) $tcnt_n = tval_p$ && $expiry_n = 0$
Scheduler:
if($expiry_p == 1$)
$task[head_p].PC = sPC_{i_p}$ && $task[head_p].REG = sReg_{i_p}$
$head_n = (head_p + 1)\%$ LENGTH
$taskid_n = task[head_n].ID$
$sPC_{o_n} = task[head_n].PC$
$sREG_{o_n} = task[head_n].REG$
$tval_n = task[head_n].TIME$
$load_n = 1$
Connecting Constraints:
$pREG_{i_n} = sREG_{o_n}$
$pPC_{i_n} = sPC_{o_n}$
$sREG_{i_n} = pREG_{o_n}$
$sPC_{i_n} = pPC_{o_n}$

Figure 9.19 Constraints generated for the system model of the processor.

9.4.2 Architecture Model

Vespa Architecture Model

The architecture of Vespa consists of 32 General Purpose Registers (GPRs) (R_0 - R_{31}), a flag register (FLAG) that has five flag bits, namely the carry bit (C), the overflow bit (V), the sign bit (N), the zero bit (Z), and the timer set bit (T). It also has a timer count register (TC) and 32 special-purpose scratch registers (S_0 - S_{31}). The T bit and the TC register are used by the processor to start the hardware timer and its expiry time, respectively. When the timer expires, it saves the current value of PC in the EPC register and jumps to the address stored in the Interrupt Address (IA) register. Vespa has a set of 30 RISC instructions detailed in [3], which were modeled in MMV. The complete behavior of the **add** instruction modeled in MMV's architectural aspect and the constraints generated by the CSP-specific code generator is provided for the purpose of illustration.

The pseudocode of the behavior of *add* instruction is shown in Figure 9.20. The source operands and destination operands of *add* are *src1, src2,* and *dest*, respectively. The first statement in the pseudocode performs the *add* operation and it is followed by statements that calculate the value for the flag bits based on the result of the addition.

1. $dest = src_1 + src_2$
2. $carryflag = ((((src_1 \,\&\, src_2) | (src_1 \,\&\, !dest) |$
$(src_2 \,\&\, !dest)) \,\&\, 2^{BitSize-1}) >> BitSize - 1)$
3. $zeroflag = (dest == 0)$
4. $signflag = dest >> BitSize - 1$
5. $overflowflag = ((((src_1 \,\&\, !src_2 \,\&\, !dest) |$
$(!src_1 \,\&\, src_2 \,\&\, dest)) \,\&\, 2^{BitSize-1}) >> BitSize - 1)$

Figure 9.20 Pseudocode for the behavior of *add* instruction.

The textual representation of the model generated for *add* in MMV is shown in Figure 9.21. Each statement in the model can be one of the following four types of statements: bitwise, shift, arithmetic, and assignment. Expression with multiple operators are represented in the model by first evaluating the result of each operator and then concatenating them using statement labels.

Constraints for ADD Instruction

The CSP generated by MMV for *add* instruction is shown in Figure 9.22. The generated CSP is a direct translation of the behavior *add* instruction. Variables with

```
Arithmetic Statement - S0  : src2 + src1
//Carryflag Behavior
BitWise Statement    - S1  : src1 & src2
BitWise Statement    - S2  : src1 & !S0
BitWise Statement    - S3  : src2 & !S0
BitWise Statement    - S4  : S1 | S2
BitWise Statement    - S5  : S4 | S3
Arithmetic Statement - S6  : BitSize - 1
Arithmetic Statement - S7  : 2^S6
Arithmetic Statement - S8  : S5 & S6
Shift Statement      - S9  : S8 >> S6
Assignment Statement - S10 : carryflag = S9
//ZeroFlag Behavior
Logical Statement    - S11 : S0 == 0
Assignment Statement - S12 : zeroflag = S11
//SignFlag Behavior
Shift Statement      - S13 : S0 >> S6
Assignment           - S14 : zeroflag = S13
//OverFlow Behavior
BitWise Statement    - S15 : S2 & !src2
BitWise Statement    - S16 : S1 & S0
BitWise Statement    - S17 : S15 | S16
BitWise Statement    - S18 : S17 & S7
Shift Statement      - S19 : S18 >> S6
Assignment Statement - S20 : overflowflag = S19
//Updating the results
Assignment Statement - S21 : dest = S0
```

Figure 9.21 Architecture level model of the *add* instruction.

subscript p and n hold the value of the GPRs and the flag bits before and after the execution of *add*. Complex instructions have conditional statements in their behavior. The code generator will generate an *if* constraint for every conditional statement in instructions behavior. CSPs for multiple instructions can be generated by assigning the output variables of one instruction as the input variable to the next instruction in sequence.

1. $dest_n = src1_n + src2_n$
2. $c_n = ((((src1_n \ \& \ src2_n) \mid (src1_n \ \& \ !dest_n) \mid (src2_n \ \& \ !dest_n)) \ \& \ 2^{31}) >> 31)$
3. $z_n = (dest_n == 0)$
4. $s_n = dest_n >> 31$
5. $v_n = ((((src1_n \ \& \ !src2_n \ \& \ !dest_n) \mid (!src1_n \ \& \ src2_n \ \& \ dest_n)) \ \& \ 2^{31}) >> 31)$

Figure 9.22 Constraints generated for ADD in MMV.

9.4.3 Refinement of Real-Time Software Scheduler

State machines for software components in the system level abstraction of the processor are manually refined into instruction sequences in the architecture abstraction. To ensure the correctness of the refinement process, the modeler marks the entities that need to be traced across the refinement. It serves as a map between the variables in the state machine model and the architecture state (i.e., registers, flag bits, memory addresses) that are equivalent in the two models. During compilation, test cases are generated and executed on the two models and the traces of the equivalent variables (marked by the modeler to be traced) in the two models are compared for any mismatches.

The system model of Vespa is refined to the architecture abstraction to demonstrate the refinement and mapping process. The programmable timer and the processor in the system abstraction refine to hardware and the ISA model in the architecture abstraction. The scheduler refines to a sequence of assembly instructions as shown in Figure 9.23, which communicates with the hardware timer via the flag bit T and register TC.

The data structure that holds the state of the scheduled processes in the process information table is stored in the memory. Each record in the process information table is 140 bytes long and the contents of each record is shown in Figure 9.24. It consists of a unique id (P_{id}) that identifies the process, the program counter value (PC) from where the process should start execution, the execution time required by the process ($Time$), and the value of the 32 GPRs and flag register. In the implementation of the scheduler, scratch registers $S0$, $S1$, and $S2$ store the memory address of the first process, current process, and the last process. These registers are assumed not to be used by the scheduled processes. The timer is a hardware unit and is manipulated by the scheduler using the flag bit T and register TC. When the timer expires, it stores the PC of the current process in EPC register and loads the PC with the start address of the scheduler. The scheduler performs the following tasks: (1) stores the architecture state of the current process to the process information table in memory, (2) increments to point at the next record, (3) loads the GPRs and Flag bits of the process with the values of the next process from the memory, (4) sets the timer count register TC with the time of the next process and sets the T bit to 1, and (5) loads the PC of the processor with the PC of the next process from memory. From the next instruction onwards, the processor will start executing the next process.

During the refinement process, the values of TC and P_{id} in the architecture model should be equal to the values stored in the variables $tval$ and $taskid$ in

```
// Store the PC of the interrupted process
start:
stx EPC, S0, #4
// Store the GPRs of the interrupted process
stx R0, S0, #12
stx R1, S0, #14
....
....
stx R31, S0, #136
stx FLAG, S0, #140
// Check to see if we have reached the end of the list
cmp S0, S1
jnz load
// Reload the start address in S0
mov S0, S2
// Load the GPR of the next process
load:
ldx R0, S0, #8
ldx R1, S0, #12
....
....
ldx R31, S0, #136
ldx FLAG, S0, #140
// Load TC with the time for the next process
ldx TC, S0, #4
// Set the TC bit to 1 to start the timer
OR  FLAG, FLAG, #10
// Load the PC with the address of the next process
ldx PC, S0, #0
```

Figure 9.23 Instruction sequence of scheduler in the architectural abstraction.

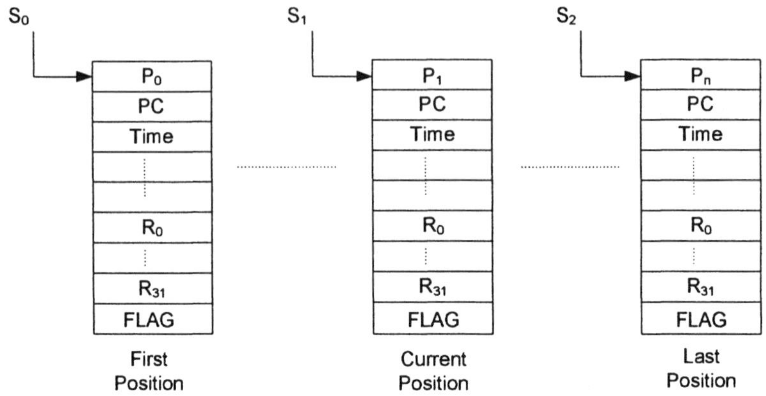

Figure 9.24 Memory map of scheduler in the architectural abstraction.

the system model. The constraints specified are $tval = TC$ and $taskid = P_{id}$. The tagged variables are traced against the marked registers and checked to ensure consistency. The code generator converts these user-specified constraints into CSP constraints that would be used to generate test cases that check for equivalence.

9.4.4 Microarchitecture Model

Vespa Microarchitecture Model

Vespa is modeled as a scalar, single-issue, five-stage pipelined processor in the microarchitecture level of abstraction [3]. The microarchitecture consists of five pipeline stages, pipeline registers, data forwarding logic, and ports to access memory. The ports are visible only at this level of abstraction. Figure 9.25 shows the high level view of the five stages in the pipeline, the pipeline registers, and the forwarded data in MMV. The I_Address, Instruction, D_Address, write_data, and read_data correspond to ports used to access memory in different stages of the processor. These ports are visible only in this level of abstraction. The register file in the microarchitecture abstraction refers to the register set defined in the architecture abstraction. All registers defined in the architecture abstraction should be mapped to a physical address in the register file at the microarchitecture abstraction. This OCL constraint was added in MMV to maintain consistency across the abstraction levels. We only show the model of IF stage and its CSP generated by the code generator. The behavior of the other stages of Vespa can be found in [3].

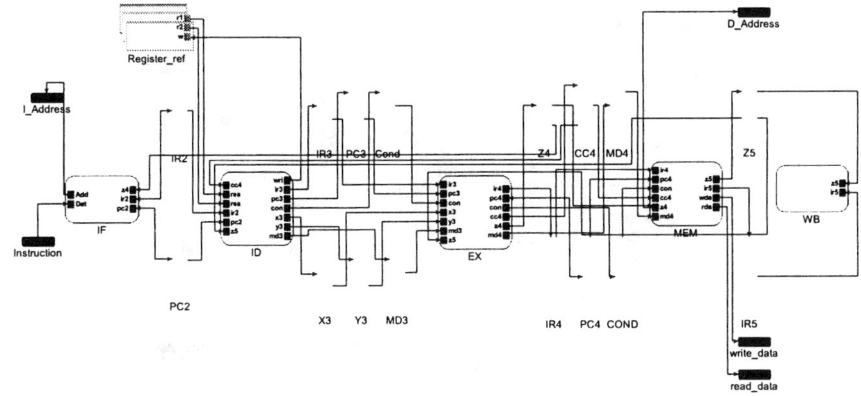

Figure 9.25 The Vespa pipeline in MMV.

Figure 9.26 shows the internals of the IF pipe stage. IF stage communicates with the ID stage by updating IR2 and PC2 registers, respectively. The address of the current instruction is stored in PC. During normal operation, the IF stage fetches the current instruction from the memory via the port *Addr* and *Data* using the address stored in PC. It then increments the PC by four to point to the next instruction in sequence. The multiplexers IRMUX, PC2MUX, and PCMUX control the value that updates the $IR2$, $PC2$, and PC registers, respectively. The control signals for the multiplexers are generated by a hazard detection and control logic in the processor. $IR2$ can either be updated with the instruction fetched from the memory, a NOP, or the recirculated value of the current $IR2$ in case of a pipeline stall. Similarly $PC2$ can either be updated with $PC + 4$ or the current value of $PC2$ depending on whether the pipeline is stalled or not. PC can be updated with $PC + 4$, the current value of PC, or the branch target address in port $z4$. The value in $z4$ is calculated in the EX stage of the processor.

Constraints for IF Stage

The CSP is generated for one time frame and CSPs for multiple time frames are generated similar to the previous two abstractions. The CSP of the IF stage is shown in Figure 9.27. The code generator formulates an *if* constraint for every multiplexer in the model. It represents the control input to the multiplexers using an integer variable. $ir2mux$, $pc2mux$, and $pcmux$ are integer variables that control the operation of the IR2MUX, PC2MUX, and PCMUX, respectively.

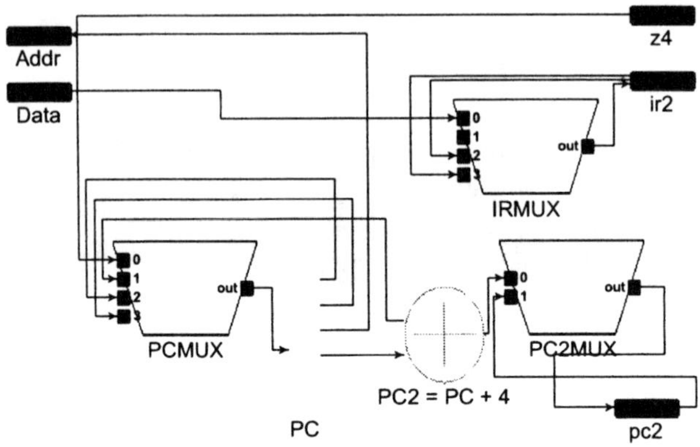

Figure 9.26 The model of the instruction fetch stage of Vespa processor.

$$\begin{aligned}
&\text{if}(ir2mux_p == 0)\; ir2_n = Mem[pc_n] \\
&\text{if}(ir2mux_p == 1)\; ir2_n = NOP \\
&\text{if}(ir2mux_p > 1)\; ir2_n = ir2_n \\
&\text{if}(pc2mux_p == 0)\; pc2_n = pc_n + 4 \\
&\text{if}(pcmux_p == 0)\; pc_+ = z4_n \\
&\text{if}(pcmux_p == 1)\; pc_+ = pc_n + 4 \\
&\text{if}(pcmux_p > 1)\; pc_+ = pc_n \\
&\text{if}(pc2mux_p > 0)\; pc2_+ = pc2_n
\end{aligned}$$

Figure 9.27 Constraints generated for IF stage.

9.5 SUMMARY

In this chapter, we proposed a microprocessor validation environment based on a metamodeling framework. The environment supports the creation of models at various abstraction levels, and keeping models at neighboring levels of abstraction consistent. This is an important feature for validation, along with the requirement that this consistent set of models be utilized for generating important validation collaterals.

We illustrate the architectural abstraction of MMV for the purpose of generating an extensible functional simulator. We illustrate how the metamodeling capability in MMV is employed to facilitate a high-level modeling of the processor architecture. On the other hand, the framework also captures the architectural specifics and the programming artifacts in order to tune the CG to generate an executable

C++ model. This clean distinction between the architectural concepts and target specifics allow for an extensible approach, which is one of the ways to tackle the modeling challenges of next generation processors.

Currently, we employ the ISS APIs for the CG, because we illustrate our approach by generating existing instructions. However as future work, we will also generate the states of the processor as well as the necessary APIs from the architectural model. One of the drawbacks with our approach is that the CG introduces many temporary variables as a side effect of the uniformity in the generation. The extra variables can be easily optimized through the compiler. On the other hand, the uniformity in the CG provides enhanced debugging and error checking capability along with highly structured and readable code.

We also demonstrated the concept by modeling a 32-bit RISC processor, Vespa, at the system, instruction set architecture, and microarchitecture levels. We showed how consistency across levels is enforced during modeling and gave an example of generating constraints from the model for automatic test generation.

References

[1] S. S. Muchnick, *Advanced Compiler Design and Implementation*, San Francisco, CA: Morgan Kaufmann Publishers Inc., 1997.

[2] F. Fallah, P. Ashar, and S. Devadas, "Simulation Vector Generation from HDL Descriptions for Observability-Enhanced Statement Coverage," *Proceedings of the Design Automation Conference (DAC'99)*, 1999, pp. 666–671.

[3] D. J. Lilja and S. S. Sapatnekar, *Designing Digital Comptuer Systems with Verilog*, Cambridge, U.K.: Cambridge University Press, 2005.

[4] C. J. Fidge, "Real-Time Scheduling Theory," Software Verification Research Center, School of Information Technology, The University of Queensland, Tech. Rep. 02-19, April 2002.

Chapter 10

Design Fault Directed Test Generation

Simulation is widely used to validate large systems like microprocessors. In simulation-based validation, a test is executed in a golden reference model as well as in the design under test (DUT), which in this case is the RTL implementation of the microprocessor. The success of simulation-based validation depends on the quality of tests. A potential fault in the RTL is exposed only if the test resulted in a state deviation of the RTL from that predicted by the golden model. Coverage metrics inspired from software validation (statement, branch, and path coverage), state machine representation (state, transition, and path coverage), and functionality of the microprocessor is used to measure the quality of tests. Effectiveness of coverage directed test generation depends on the *types of faults* that can occur and the strength of the *coverage metric* to detect these faults.

We analyze the bug descriptions and modeling errors reported during the study of microprocessor validation in [1, 2] and relate them to the fault types observed during software validation [3]. The outcome is a high correlation between the fault types and the modeling errors seen in the control logic of microprocessors. Hence, we propose a model-based test generation framework that detects these fault classes.

The framework allows modeling the microprocessor at the architectural and microarchitectural abstraction. The modeling language provides the syntax and semantic necessary to describe the architectural and microarchitectural models. The framework translates these models into constraint satisfaction problems [4] that are resolved through a highly scalable constraint solver. The architectural model is converted into architectural constraints that are solved to generate random test suites that validates the *microcode*. The microarchitecture model is converted into microarchitectural constraints that are solved to generate random test suites that validate the RTL implementation. However, to improve the quality of the test suite generated, we enable our framework with coverage-directed test generation capability. The coverage metric is used to generate additional constraints that are given to the solver along with the constraints obtained from the model. The validation flow is shown in Figure 10.1.

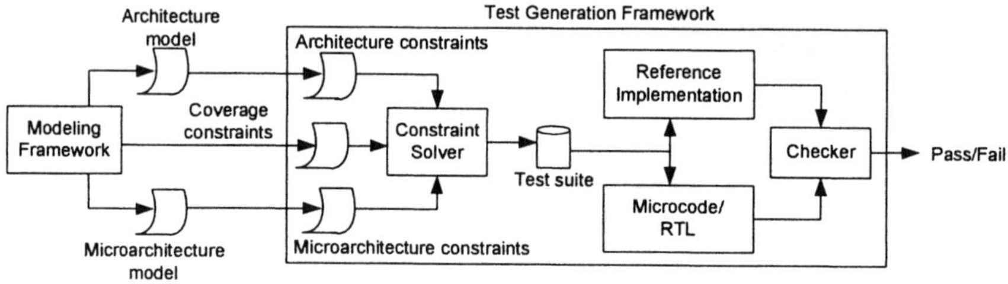

Figure 10.1 Validation flow.

We support traditional coverages such as statement and branch as well as modified condition decision coverage and design fault coverage. Modified condition decision coverage (MCDC) [5] is a coverage metric that is widely used for the validation of critical software systems like avionics. The test suite generated as a result of MCDC-directed test generation is found to detect most of the design fault classes proposed in [3]. Therefore, we propose MCDC as a metric for microprocessor validation. The design fault classes not covered as result of MCDC are detected by generating additional constraints that tune the solver to generate solutions that test them. We illustrate our test generation framework by creating test suites for Vespa [6], a 32-bit pipelined microprocessor. We provide results to highlight the importance of MCDC-directed test generation for design fault coverage.

10.1 MOTIVATION

Velev [2] and Campenhout [1] conducted empirical studies on error/bug categories observed in academic microprocessor projects. The bug categories identified were very similar to the faults observed during validation of software systems [5]. The faults were analyzed and an extended fault model comprising of nine types of faults on Boolean expressions was proposed by Lau et al. shown in Table 2.2. Being able to detect these basic fault classes would result in identifying more complex faults, which follows from the *fault coupling effect hypothesis* [3]. We correlate the bug categories in [1, 2] to the extended fault model in [3] and motivate the need for a test generation framework that provides coverage for the fault classes.

In [1], Campenhout et al. analyzed modeling errors commonly seen in microprocessors and generated error distributions. We segregated the errors related to signals and gates in the control logic of the processor into three groups: (1) wrong, (2) missing, and (3) extra. The study reported 33% of the errors occur due to wrong signal source and 4.9% due to wrong gate type. The missing gate errors accounted for 7.4% and missing input(s) for 11.5%. The errors from extra inversion and extra insertion contributed 1.6% each. We relate these errors to the fault classes based on omission, insertion, and incorrect reference (shown in Table 2.2) and summarize the correlation in Table 10.1. Note that the exact description of the modeling error was not provided and certain assumptions were made to arrive at the correlation.

Table 10.1

Correlation Study

Fault Class in [3]	*Campenhout et al. [1]*	*Velev [2]*
ENF, LNF	1.6% (extra)	-
TOF	7.4% (missing)	56.8% (159/280)
LOF	11.5% (missing)	11.4% (32/280)
LIF	1.6	-
LRF	32.8% (wrong)	26.4% (74/280)
ORF	4.9% (wrong)	0.4% (1/280)

Velev [2] studied three microprocessors to perform a bug collection in pipelined and superscaler designs and arrived at the following bug categories: (1) incorrect control logic, (2) incorrect connections, (3) incorrect design for formal verification, and (4) incorrect symbolic simulation. His work identified 280 bugs and presented a neat description of each bug, which was subjected to our analysis. We found that 95% of the bugs identified were subsumpted by the fault classification in [3] and the remaining 5% were artifacts of the formal verification framework in [2].

10.2 MODELING AND TEST GENERATION

The modeling framework facilitates microprocessor modeling through an abstract representation called the metamodel discussed in Chapter 9 [7]. The metamodel defines the syntax and semantic for the language in which the modeler describes the architecture and microarchitecture of the processor.

10.2.1 Architecture Modeling

The processor is specified through a register/memory-level description (registers, their whole-part relationship, and memory schematic) and an instruction set capture (instruction-behavior). The metamodel for architectural modeling provides the user with constructs to describe function blocks, if-else blocks, statement sequences, and loops as well as the register map and memory layout. On the left of Figure 10.2, we show the jump-if-not-zero instruction (JNZ) that positions the instruction pointer (IP) to the location specified in source (src) when the zero flag (ZF) is not set. The IP and ZF are references to registers modeled as a part of the architecture. The register map specifies all the registers as well as their whole-part relationship as shown on the right in Figure 10.2, whereas the memory schematic describes the memory hierarchy. The *src* is an identifier that translates to a register/memory reference or an immediate value during decoding.

Function block: JNZ
Identifier: src
Register reference: IP
Register reference: ZF
If block:
Guard:
Relational Statement: ZF = 0
Action:
Arithmetic Statement: IP = IP + src

Pentium® IV
Register: RAX [63:0]
Reg reference: EAX [31:0]→ RAX
Reg reference: AX [15:0] → EAX
Reg reference: AH [15:8] → AX
Reg reference: AL [7:0] → AX

Figure 10.2 Architectural modeling.

10.2.2 Microarchitecture Modeling

The language allows the modeler to instantiate and configure different pipe stages, provide ports to interface with the memory and register file, insert instruction registers, and describe the control logic needed to perform data forwarding and stalling. It provides a set of basic configurable components such as MUX_n, ALU, and INC_n, which can be combined to create complex stages that interact with each other as well as with the architectural entities through access ports. In Figure 10.3, we show the update of the program counter in the instruction fetch stage (IF) of the Vespa processor [6] modeled using our constructs. The selection criteria of the MUXes are not shown, since they map to statements similar to the ones in architectural modeling. PC and PC2 are program counters, where PC belongs to IF and PC2 is an input to instruction decode (ID) stage. PC2 is updated with the PC or previous PC2 value depending on whether the pipeline is stalled or not. PC is updated with PC+4 (location of the next instruction), the previous PC/PC2 (pipeline stall), or a new address at port Z (branch instruction). The value at Z is obtained from the EX stage.

//PC Multiplexing
$\mathbf{MUX_4}$: PC_MUX
$\mathbf{PC\_MUX_{ip}}$: Z, PC, PC+4, PC2
$\mathbf{PC\_MUX_{op}}$: PC
//Update PC2 after incrementing PC
$\mathbf{INC_4}$: INC_PC
$\mathbf{INC\_PC_{ip}}$: PC
$\mathbf{INC\_PC_{op}}$: PC2
//PC2 Multiplexing
$\mathbf{MUX_2}$: PC2_MUX
$\mathbf{PC2\_MUX_{ip}}$: PC, PC2
$\mathbf{PC2\_MUX_{op}}$: PC2

Figure 10.3 PC behavior of the IF in Vespa [6].

Metamodel-based modeling is very similar to having a library of configurable components that the modeler instantiates and characterizes to perform a certain functionality. Such a language need not be highly expressive, but it should be extendable. The customizability obtained from the metamodel makes the modeling framework easily extendable with specialized components needed for speculative and out-of-order execution such as reorder buffers and memory arrays.

The test generation framework as shown in Figure 10.4 converts the architectural and microarchitectural model into constraint satisfaction problems (CSPs) [4] and passes them through a constraint solver and a test case generator to create test suites that validate the microcode and RTL. In order to correctly formulate the CSP, we convert the architectural and microarchitectural model into flow graphs and generate constraints in the form of single assignments.

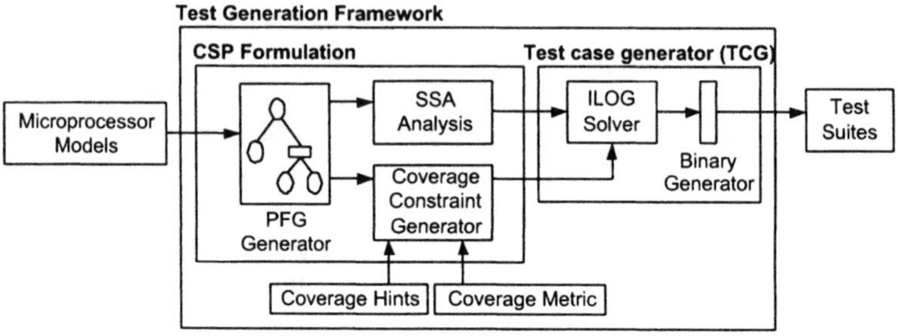

Figure 10.4 Test generation framework.

10.2.3 CSP Formulation

A model created using our modeling framework is converted into a program flow graph (PFG), where all the architectural/microarchitectural constructs are mapped to a bunch of states, decisions, and transitions. For an architectural model, the statements translate to states and the if-else constructs map to decisions which capture the guard that causes a transition. Function blocks map to hierarchical states, whereas loops lead to cyclic graphs. The identifiers, registers, memory locations, and register references are used to instantiate variables that undergo a change in value whenever a statement updates them. In the microarchitecture model, the basic components translate to a bunch of if-else sequences that when evaluated to *true* execute a sequence of statements. For example, the MUX is basically an *If block* sequence with *Assignment Statements* that execute based on which selection criterion is valid in that run. The PFG for the JNZ instruction and the PC behavior of the fetch stage of Vespa is shown in Figure 10.5.

For a graph with a unique path (no branches), the behavior is converted into CSP in a straightforward manner by generating constraint variables and assignment constraints. However for a model with multiple paths, the graph generated undergoes static single-assignment (SSA) analysis [8] to construct the correct set of constraints for the CSP formulation for one complete run (entry to exit). The analysis focuses on states with multiple incoming transitions, which maps to a statement that follows a set of conditional branches. At these join states, parent resolution and variable renaming are performed very similar to how *dominance frontier* [8] is computed in SSA during compiler analysis. The outcome of the analysis is variables and decisions being inserted to help in the CSP formulation. The resulting variables from the SSA analysis for the PC behavior in the IF stage are shown on the left in Figure 10.6, where ($D_i \implies S_i$) means if D_i is true then S_i is executed.

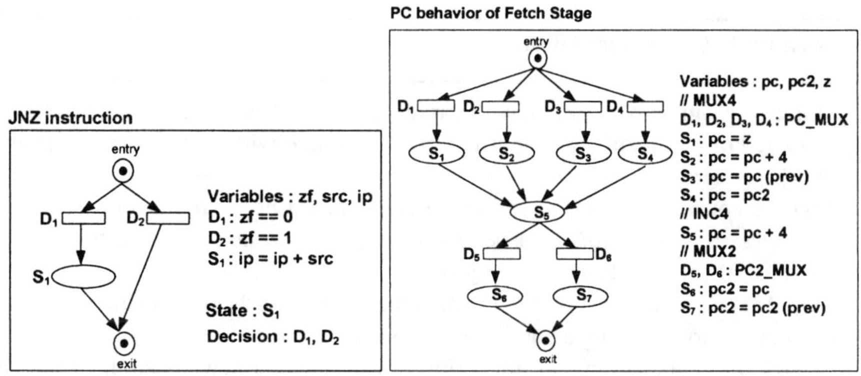

(a) Instruction model. (b) Pipeline stage model.

Figure 10.5 (a, b) Program flow graph.

Var $pc_0, pc_1, pc_2, pc2_0, pc2_1, z_0$
$D_1 \implies S_1\colon pc_1 = z_0$
$D_2 \implies S_2\colon pc_1 = pc_0 + 4$
$D_3 \implies S_3\colon pc_1 = pc_0$
$D_4 \implies S_4\colon pc_1 = pc2_0$
$S_5\colon pc_2 = pc_1 + 4$
$D_5 \implies S_6\colon pc2_1 = pc_2$
$D_6 \implies S_7\colon pc2_1 = pc2_0$

Var $pc_0, pc_1, pc_2, pc_3, pc2_0, pc2_1, z_0$
$D_1 \implies S_1\colon pc_1 = z_0$
$D_2 \implies S_2\colon pc_1 = pc_0 + 4$
$D_3 \implies pc_0$
$D_4 \implies S_4\colon pc_1 = pc2_0$
Additional Decisions
$D_1 \vee D_2 \vee D_4 \implies S_8\colon pc_2 = pc_1$
$D_3 \implies S_9\colon pc_2 = pc_0$
$S_5\colon pc_3 = pc_2 + 4$
$D_5 \implies S_6\colon pc2_1 = pc_3$
$D_6 \implies S_7\colon pc2_1 = pc2_0$

Figure 10.6 SSA analysis outcome of the PFG in Figure 10.5(b).

Consider the decision D_3 and state S_3, which does not change the current value of pc but forwards the previous value ($pc_1 = pc_0$). Omission of S_3 results in a contention at S_5 with multiple paths, where most of them update the *pc* register except for one (S_3). Therefore, a parent resolution is performed that inserts additional decisions and variables as shown on the right in Figure 10.6. The SSA is nontrivial when considering nested-ifs and loop unrolling. The CSP formulation guidelines are outlined here:

Algorithm 1: CSP Formulation

{Given a program flow graph, with variables, states, decisions, and transitions}
Step 1: For a **variable**, an input constraint variable with a finite domain [min, max] is generated.
Step 2: For a **decision**, the following state should be exercised only when it evaluates to *true* or *false*, respectively, and the decision is converted into a conditional constraint of the form *if-then*.
Step 3: For a **state**, a set of input constraint variables is used to update an output constraint variable using an assignment constraint (=).
Step 4: If a state assigns **a new value** to a variable, a new constraint variable is generated for that variable (renaming act in SSA) and future references to the variable will refer to the newly generated constraint variable.
Step 5: If a state assigns **multiple new values** for a program variable, a new constraint variable is generated and for each decision that results in a value, the corresponding conditional constraint is generated such that the new variable can be assigned that value (parent resolution in SSA).

The language used to print out the CSP consists of assignment and conditional constraints. The CSPs developed for the above examples are shown in Figure 10.7. On the left, the architectural constraints for the CSP generated from the JNZ instruction are shown and on the right the microarchitectural constraints for the CSP generated from the PC behavior in the IF stage are shown.

10.2.4 Test Case Generator (TCG)

The CSP is given as input to the constraint solver in the TCG. The constraint solver used in our framework is ILOG, which is a commercial solver designed for performance and scalability. The solution generated by the solver consists of a possible value for **every constraint variable** in the problem such that the CSP is satisfied. This solution is called a *test case*. Examples of some possible solutions generated for Figure 10.7 are shown in Figure 10.8. Test case T_1 causes

if($zf_0 = 0$) then $ip_1 = src_0 + ip_0$
else $ip_1 = ip_0$

if($D_1 = 1$) then $pc_1 = z_0$
else if($D_2 = 1$) then $pc_1 = pc_0 + 4$
else if($D_4 = 1$) then S_4: $pc_1 = pc2_0$
Step 5 Illustration for pc_2
if(($D_1 \vee D_2 \vee D_4$) = 1) then $pc_2 = pc_1$
else if ($D_3 = 1$) then $pc_2 = pc_0$
$pc_3 = pc_2 + 4$
if($D_5 = 1$) then $pc2_1 = pc_3$
else if($D_6 = 1$) then $pc2_1 = pc2_0$

Figure 10.7 CSP formulation for Figures 10.5(a) and 10.6 (right side).

the instruction to jump to the location src, whereas T_2 does not. Test case T_3 causes the IF stage to fetch a branch instruction from #1100 and T_4 causes both IF and ID stages of the pipeline to be stalled.

Test cases for JNZ instruction
$T_1 = \{zf_0 = 0, src_0 = 1001, ip_0 = 5, ip_1 = 1006\}$
$T_2 = \{zf_0 = 1, src_0 = 1001, ip_0 = 5, ip_1 = 5\}$
Test cases for PC behavior in IF
$T_3 = \{pc_0 = 1000, pc2_0 = 996, z_0 = 1100, pc_1 = 1100\ (D_1 = 1), pc_2 = 1100, pc_3 = 1104, pc2_1 = 1104\}$
$T_4 = \{pc_0 = 1000, pc2_0 = 996, z_0 = 1100, pc_1 = 1000\ (D_3 = 1), pc_2 = 1000, pc_3 = 1004, pc2_1 = 996\}$

Figure 10.8 Possible test cases generated for Figure 10.7.

The binary generator creates a program binary by combining the model with the registers and memory locations initialized to the values stored in the input constraint variables obtained from the test case. A program binary for the PC's behavior initializes Z, PC, and PC2 to the values of z_0, pc_0, and $pc2_0$, respectively. The output constraints pc_3 and $pc2_1$ are the results obtained by executing the test case against the reference model.

10.2.5 Usage Mode

There are two modes of usage for the test generation framework: (1) model-random and (2) coverage-directed. A test suite in the model-random mode is generated by converting the model into a CSP and solving it multiple times through the solver. Attaining a coverage goal in this mode requires a large number of test cases. However, validators are interested in design-fault based testing, the test cases for which are difficult to obtain using the model-random approach. As a result, the framework also provides a coverage-directed approach that generates one possible test suite, which attains 100% coverage of a specific goal. In this mode, coverage constraints are given as input to the solver that directs the test generation towards the goal as shown in Figure 10.4.

10.3 COVERAGE CONSTRAINTS

The coverage constraint generator (CCG) takes two sets of inputs besides the program flow graph: First, the coverage annotations that the modeler provides to expedite the reachability of the coverage goal and second, the coverage type based on which the additional constraints are generated. The coverage hints are embedded into the flow graph for certain coverage types and for the others specialized constructs that serve as collectors are provided. The annotations are done by setting the attribute **MARK** associated with a state and decision. The coverage types supported are: (1) statement coverage, (2) branch coverage, (3) MCDC, and (4) design fault coverage.

10.3.1 Statement Coverage

To perform statement coverage, every statement in the model description needs to be reached at least once. The CCG works off the flow graph of the model and generates the additional constraints needed to enforce that all the states are executed. The validator is allowed to mark the interesting states of the graph as hints for the CCG. The marked states are collected, ordered based on their depth, and a set of constraints is generated that indicates a path to each of these states within the model. For example in the PFG shown in Figure 10.5(a), $O(D_1) = O(D_2) = 1$ and $O(S_1) = 2$. If none of the states are marked, then all the states are collected, ordered, and then a longest path algorithm is run until each state is covered at least once. For each of these paths the corresponding constraints are generated and fed to the solver to create a test case that is part of the test suite for statement-directed test generation. Running the CCG in either mode will generate the same test suite.

10.3.2 Branch Coverage

To perform branch coverage, every if-else block in the model description needs to be executed when it evaluates to *true* as well as *false*. The CCG works off the flow graph and generates the conditional constraints needed to enforce every decision is evaluated to *true* and *false*. The validator is allowed to tag the problematic decision points and the CCG collects and orders them before generating the constraints. For the n decisions collected and ordered, the CCG generates $2n$ test cases. Duplication of test cases is a side effect of this approach for the implicit decisions that evaluate to *true* or *false* due to explicit setting of some lower-order decisions.

Function 1: branch_cov(d, val)

{Given a decision d, the truth value *val*, and the additional constraint set $C_C = \emptyset$
cr_constraint(t) - creates a constraint that enforces t.}
Step 1 C_C = **cr_constraint**($d = val$)
Step 2 return C_C

For every collected decision, the CCG identifies a path that reaches it and generates the constraints for the path as well as the two conditional constraints for its evaluation. Generating the constraints for a path may result in setting the decision along the path. We adopt a heuristic that begins with the lowest ordered decision in the collection and moves down the flow graph.

Algorithm 2: MCDC Constraint Generation

{Given a program flow graph G, D = **collect$_d$**(G) and $T_S = \emptyset$
collect$_d$(G) - gets all decisions in G with the MARK attribute set.
collect$_c$(d) - gets all conditions in decision d.
arrange$_d$(D) - sorts the decision list D in ascending order based on depth.
find_path(G,d) - gets a path from the root node *entry* in G to the node d.
gen_constraint(P) - generates the constraints for the path P.
check_boolean_opr(d) - checks if d is a *condition* with no Boolean operators or a *decision* with one or more operators.}
Step 1 $\forall\ d \in$ **arrange$_d$(D)**
  Step 1.1 P = **find_path**(G, d)
  Step 1.2 if **check_boolean_opr**(d) == *false* then
    T_T = **gen_constraint**(P) ∪ **cr_constraint**($d = 1$)

T_F = **gen_constraint**(P) ∪ **cr_constraint**($d = 0$)
$T_S = T_T \cup T_F$
Step 1.3 else
Step 1.3.1 $\forall\, c \in$ **collect$_c$**(d)
T_T = **gen_constraint(P)** ∪ **MCDC_cov**(c, d, *true*)
T_F = **gen_constraint(P)** ∪ **MCDC_cov**(c, d, *false*)
$T_S = T_T \cup T_F$ and $T_T = T_F = \emptyset$
Step 1.4 rm_duplicate(T_S)

10.3.3 MCDC

The objective of MCDC is that every condition within a decision should independently affect the outcome of the decision. The CCG works off the flow graph and generates the constraints necessary to attain the objective using Algorithm 2. The validator is allowed to tag the problematic decision points and the CCG collects them using **collect$_d$**. It then orders these collected nodes using **order$_d$** based on their depth in the graph before generating the constraints. For every collected decision, the CCG identifies a path that reaches it using **find_path** and generates the constraints for the path as well as the additional MCDC constraints for that decision. The constraints for the path is generated using the **gen_constraint** function. For a decision that is just a condition, branch and MCDC generate identical test suites. However for a decision with one or more Boolean operators found using **check_boolean_opr**, the additional constraints are generated to tune the evaluation of the decision to the condition's effect. For each condition, the decision should evaluate to *true* and *false* based on whether it is assigned a "0" or a "1." This results in m test cases for the *0-value* scenario and another m for the *1-value* scenario. After removal of the duplicate test cases using **rm_duplicate** MCDC results in $m + 1$ test cases. To perform MCDC on a decision d with m conditions, with respect to the jth condition, CCG generates constraints for both the *0-* and *1-value* scenarios using Function 1.

Function 2: MCDC_cov(c_s, d, val)

{Given a condition c_s in decision d and *val*-scenario and the additional constraint set $C_C = \emptyset$ }
Step 1 d' = rename(copy(d))
Step 2 C_C ∪ **cr_constraint**($d = val \wedge d' = !val$)

Step 3 C_d = **collect$_c$**(d) and $C_{d'}$ = **collect$_c$**(d')
Step 4 $\forall\ c_i \in C_d$
  Step 4.1 if $c_i = c_s$ then $C_C \cup$ **cr_constraint**$(c_i <> C_{d'}[\mathrm{i}])$
  Step 4.2 else $C_C \cup$ **cr_constraint**$(c_i = C_{d'}[\mathrm{i}])$
Step 5 return C_C

Function 2 **MCDC_cov**(c_s, d, val) copies d into d' and renames the conditions in d'. Then, it generates constraints that assert d to val and d' to $!val$ by the **cr_constraint** function. It also enforces that every condition in d and d' have identical values except for the condition c_s in d and its corresponding copy in d' using the same function.

10.3.4 Design Fault Coverage

The design fault classes are provided in Table 2.2. To attain this coverage goal, test cases that cover each of these fault classes are generated by the CCG. *Our probabilistic analysis of the fault detection capability of MCDC revealed that seven of the nine fault classes are covered by the test suite generated as a result of MCDC-directed test generation.*

Consider the literal omission fault class (LOF). MCDC detects this fault because to satisfy this coverage, every literal in the decision has to affect the output of the decision. Therefore a pair of test cases that causes the missing literal to affect the decision's output in the reference model exists, which is missing in the implementation. For at least one of these test cases, the output of the decision in the implementation will differ from the fault-free decision. Therefore, MCDC-directed test generation finds the bug with a probability 1. The only fault class not covered by MCDC with a probability 1 is literal reference fault, for which CCG produces additional constraints that guide the TCG to detect it.

Example:
$d = A \wedge B \vee C$ (fault-free decision)
$d' = A \wedge B$ (LOF'y decision)
Test cases generated:
case A : [0, 1, 0] (d=0) and [1, 1, 0] (d=1)
case B : [1, 0, 0] (d=0) and [1, 1, 0] (d=1)
case C : [0, 1, 0] (d=0) and [0, 1, 1] (d=1)
Test case [0,1,1] executed on d', results in $d <> d'$

Literal Reference Fault Coverage: The annotations to attain this coverage are captured using the **LRF_list** construct, which allows the creation of a list of variables. The validator associates an LRF_list with every problematic variable v that tells CCG that v can be replaced by an instance of any variable in the given LRF_list, which results in a fault. If an LRF_list is not provided then, the CCG assumes that every variable can be replaced incorrectly by every other variable, which results in a large set of test cases. Therefore, the validator is recommended to specify a possible LRF_list with a variable and mark that variable, such that CCG can take advantage of it. The constraints generation for LRF coverage is shown in Algorithm 3.

Algorithm 3: LRF Constraint Generation

{Given a fault-free decision d with m literals in DNF, an LRF_list associated with literal $x_i \in d$, and $T_S = \emptyset$ }
Step 1 $\forall\ y_i \in$ LRF_list
  Step 1.1 d_e = copy(d, 0, i-1) $\vee\ y_i \vee$ copy(d, i+1, m)
  Step 1.2 T_F = **cr_constraint(**$d = 0 \wedge d_e = 1$)
  Step 1.3 $T_F \cup$ **cr_constraint(**$y_i <> x_i$)
  Step 1.4 T_T = **cr_constraint(**$d = 1 \wedge d_e = 0$)
  Step 1.5 $T_T \cup$ **cr_constraint(**$y_i <> x_i$)
  Step 1.6 $T_S = T_T \cup T_F$ and $T_T = T_F = \emptyset$

Literal Insertion Fault Coverage: The coverage annotations are captured with the **LIF_list** construct, which the validator associates with every problematic decision d informing the CCG that d can be inserted with a variable from the given LIF_list resulting in a fault. The constraints generation for LIF coverage is shown in Algorithm 4.

Algorithm 4: LIF Constraint Generation

{Given a fault-free decision d with m literals in DNF, an LIF_list associated with decision d, and $T_S = \emptyset$ }
Step 1 $\forall\ y_i \in$ LRF_list
  Step 1.1 $d_e = d \vee y_i$
  Step 1.2 T_F = **cr_constraint(**$d = 0 \wedge d_e = 1$)
  Step 1.3 T_T = **cr_constraint(**$d = 1 \wedge d_e = 0$)
  Step 1.4 $T_S = T_T \cup T_F$ and $T_T = T_F = \emptyset$

After specifying the coverage type and inserting the coverages hints, the CCG generates the coverage constraints. These are given as additional constraints to the TCG to generate the test suite that achieves the coverage goal.

10.4 RESULTS

We applied our test generation methodology on Vespa [6], a 32-bit pipelined RISC microprocessor to evaluate our framework. The microarchitecture of Vespa was modeled using our metamodel-based language and converted into a program flow graph from which a CSP was formulated. Additional constraints for each coverage metric were generated using the coverage constraint generator. These coverage constraints along with the CSP are resolved through a solver to generate the coverage-specific test suite. Model-random test suites were generated by asserting the statements and decisions in the model with uniform probability.

The model of the microprocessor consists of 330 unique decisions and 400 unique conditions, where the number of conditions in a decision varied between 1 and 10. These decisions translate to Boolean expressions in the HDL implementation of Vespa and are prone to modeling errors. Being able to generate tests that detect these errors improve the quality of functional validation.

To evaluate our framework, we inserted modeling errors into the model from the fault classes described in [3]. Seven modeling bugs (LOF, LNF, TOF, TNF, ENF, $ORF[+]$, and $ORF[.]$), were inserted into each of the 330 decisions in the model. For LRF, 50 decisions were randomly chosen and the variables in the decisions were randomly replaced with other variables in the model. For each inserted bug, we compute the test-set that would detect the bug. Next, we compared the test cases in the test suite generated for model-random, statement, branch, and MCDC with the set of test cases that detect each inserted bug (the intersection of the two test suite). The quality of different coverage metrics is measured by the number of distinct bugs detected by a test suite generated to satisfy a coverage metric. In order to do a fair comparison, the number of test cases in the test suites being compared was adjusted to be the same. Table 10.2 shows the percentage of bugs detected by each of the test suites.

It is seen that MCDC is uniformly better in detecting each of the design faults when compared to statement and branch. MCDC detects 7 of the 9 fault classes 100% of the time. Statement and branch are able to cover some of the fault classes completely and for some they perform badly. Therefore, these metrics are not stable and weaker than MCDC. This illustration supports our proposal to use MCDC as

Table 10.2
Fault Coverage of Different Metrics

S.No	*Fault*	*Random (%)*	*Stmt (%)*	*Branch (%)*	*MCDC (%)*
1	LOF	92.7	28.9	32.7	100
2	LNF	99.9	79.9	80.2	100
3	TOF	92	32	36	100
4	TNF	100	76.4	77.6	100
5	ENF	100	100	100	100
6	ORF[+]	96	32	36	100
7	ORF[.]	98	66.6	61.7	100
8	LRF	92.5	72.5	95	95

a coverage goal for microprocessor validation. Model-random on the other hand performs better than statement and branch because random tests generated in our framework can assert multiple decisions and statements simultaneously resulting in a higher fault coverage. However, model-random test generation fails to achieve 100% coverage of design faults such as LOF and TOF, which are commonly seen bugs in microprocessor validation as shown in Table 10.1.

10.5 SUMMARY

Simulation-based validation is widely acknowledged as a major bottleneck in current and future microprocessor design due to the increasing complexity. A possible solution is to perform design error based test generation from a high-level description of the microprocessor. Our methodology made two important contributions. First, a metamodel-based modeling framework was developed that captures the structure and behavior of the architecture (register file and ISA) and the microarchitecture (pipeline structure, control, and data forwarding logic). CSP formulation of these models resulted in test suites that covered design fault commonly seen during microprocessor validation, therefore illustrating our coverage directed test generation approach. Second, we show a high correlation between the design faults proposed by software validation and modeling errors/bugs observed during the study microprocessors validation.

Our future work includes extending the test generation framework to perform exception coverage, event coverage, and sequencing coverage for instructions. We will also perform empirical evaluation of the bug finding capability of different coverage metrics as well as the stability of different coverage metrics across test suites.

References

[1] D. V. Campenhout, T. Mudge, and J. Hayes, "Collection and Analysis of Microprocessor Design Errors," *IEEE Design and Test of Computers*, Vol. 17, No. 4, 2000, pp. 51–60.

[2] M. Velev, "Collection of High-Level Microprocessor Bugs from Formal Verification of Pipelined and Superscalar Designs," *Proceedings of the International Test Conference (ITC'03)*, 2003, pp. 138–147.

[3] M. F. Lau and Y. T. Yu, "An Extended Fault Class Hierarchy for Specification-Based Testing," *ACM Transactions on Software Engineering and Methodology*, Vol. 14, No. 3, 2005, pp. 247–276.

[4] L. Bordeaux, Y. Hamadi, and L. Zhang, "Propositional Satisfiability and Constraint Programming: A Comparative Survey," *ACM Computing Surveys (CSUR)*, Vol. 38, No. 4, 2006.

[5] K. Hayhurst, *A Practical Tutorial on Modified Condition/Decision Coverage*, National Aeronautics and Space Administration, Langley Research Center, 2001.

[6] D. J. Lilja and S. S. Sapatnekar, *Designing Digital Comptuer Systems with Verilog*, Cambridge U.K.: University Press, 2005.

[7] A. Dingankar, D. Mathaikutty, S. Kodakara, S. Shukla, and D. Lilja, "MMV: Metamodeling Based Microprocessor Validation Environment," *Proceedings of IEEE International High Level Design Validation and Test Workshop (HLDVT'06)*, November 2006, pp. 143–148.

[8] S. S. Muchnick, *Advanced Compiler Design and Implementation*, San Francisco, CA: Morgan Kaufmann Publishers Inc., 1997.

Chapter 11

Model-Driven System-Level Validation

In this chapter, we present a model-driven methodology for development and validation of system-level designs, which would further enhance the verification experience by providing protocol checking and functional coverage metrics as collaterals. In our validation flow shown in Figure 11.1, we begin with a specification document that describes the design requirements in a natural language such as English. We derive a functional model (reference model) from the specification in a language called Esterel [1]. This language easily allows abstracting away the implementation details and concentrating on the functional behaviors. Its state machine formalism renders the functional model amenable to formal verification, which we leverage for the purpose of test generation through the development environment called Esterel Studio (ES) [2].

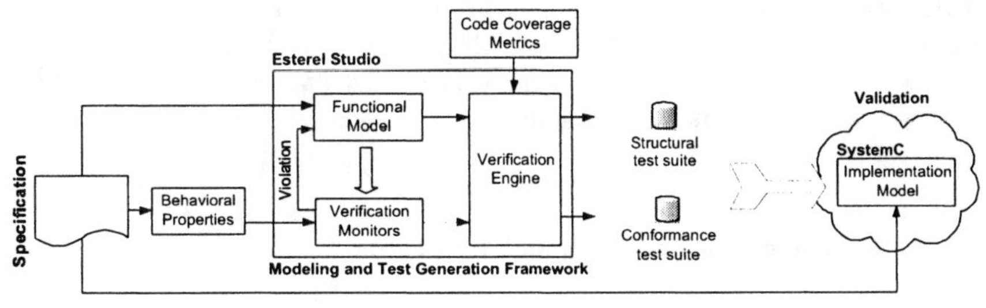

Figure 11.1 Model-driven validation flow.

Properties are extracted from the English specification and modeled as verification monitors in Esterel. These monitors are employed in: (1) verifying the functional model and (2) generating conformance tests through Esterel Studio. These conformance tests are used to attain functional coverage of the implementation model. Furthermore, we instrument the functional model with assertions to generate structural tests through Esterel Studio, which are used to attain code coverage of the implementation. These structural and conformance tests are transformed into concrete system-level tests that are utilized in the validation of the implementation model. The implementation model is also developed from the specification in the system level language SystemC [3].

11.1 TEST GENERATION METHODOLOGY

Our three-stage methodology is shown in Figure 11.2. In Stage 1, the user begins by building a *functional model* in Esterel [1]. The user also derives behavioral properties from the English specification [4] in PSL and creates a *verification model* in Esterel, which is used to verify the correctness of the functional model. In Stage 2, the functional model and verification model are instrumented using assertions for the purpose of test generation. The functional model is annotated with assertions for code coverage, which results in a *structural coverage model*. The verification model is annotated with assertions for functional coverage, which results in a *conformance coverage*[1] model. In Stage 3, these annotated models are passed through the formal verifier (*esVerify*) in ES to generate input scenarios in *.esi files*. The input scenarios are translated into testbenches in C along with the functional model using the code generation capability of ES. The testbenches are compiled and executed on the C functional model and the outputs are recorded in *.eso files*. The input and output scenarios generated (.esi and .eso files) are used to construct the abstract test suite, which is transformed into system-level tests that attain structural or functional coverage on the implementation. Note that all models developed in our integrated framework are executables.

11.1.1 Modeling and Simulation and Verification

In Stage 1, we create a functional model from the specification and a verification model from behavioral properties derived from the specification in PSL.

1. Conformance coverage is the same as functional coverage.

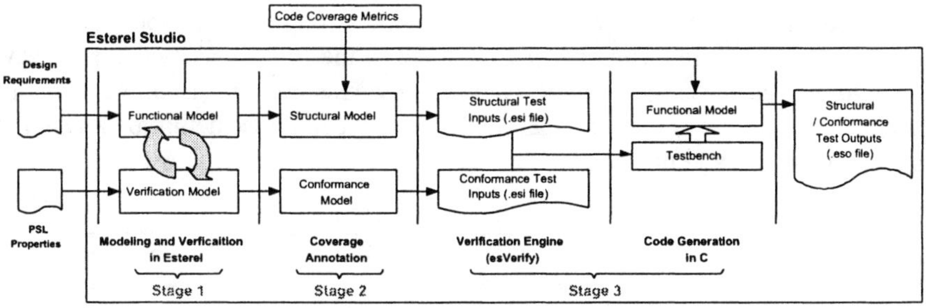

Figure 11.2 Proposed test generation methodology.

Functional Model

The functional model is a formal executable specification of the designer's intent described in Esterel. We define an Esterel-based modeling paradigm that allows the functional model to be endowed with implementation characteristics, which are exploited during test generation. Examples include the module boundaries and module instantiations, which allow for identical input-output interfaces and structural hierarchy. The control paths are also mapped identically to the implementation, but the data paths are abstracted into an array of Booleans through specialized data handling objects. We illustrate the modeling paradigm by creating a functional model in Esterel for the *fir filter* from the SystemC distribution [3].

The modeling paradigm in Esterel begins by defining the data handling objects, which describe the type definitions. We provide the modeler with data handling objects that are abstractions of datatypes in the implementation domain. For example, let us consider SystemC to be our implementation language, which has a datatype $\texttt{sc\_int}\langle N\rangle$ that is commonly used by instantiating N. The corresponding data object definition and instantiation in Esterel is shown in Listing 11.1.

Listing 11.1 Esterel data object for a SystemC datatype.

```
data sc_data :
    generic constant N : unsigned;
    type sc_int = bool[N];
end data

module A :
    extends data sc_data [constant 8 / N]; //sc_int<8> datatype
    ...
end module
```

The interface header which captures inputs, outputs, and local signals is defined next in Esterel. The interfaces defined in the functional model must have a one-to-one correspondence with component boundaries in the implementation model. For example, in Listing 11.2, we illustrate the interface definition `FirIntf` in Esterel for the *fir* component.

Listing 11.2 Esterel interface header for the *fir* component.

```
//Esterel model                              // SystemC model
interface FirIntf :                          SC_MODULE( fir ) {
  input reset          : bool;                 sc_in<bool>  reset;
  input in_valid       : bool;                 sc_in<bool>  in_valid;
  input sample         : integer;              sc_in<int>   sample;

  output out_data_rdy  : bool;                 sc_out<bool> out_data_rdy;
  output result        : integer;              sc_out<int>  result;
end interface                                  sc_in_clk    CLK;
                                               ... }
```

The *CLK* input to the fir component is not explicitly provided as a part of the interface header, because the notion of a single clock (tick) is implicit in the Esterel programming paradigm. The fir component in SystemC is modeled as an `SC_THREAD`, which is triggered by the *CLK* input. The *fir* constitutes a reset behavior and the filter computation when inputs are valid. The equivalent description in Esterel is shown in Listing 11.3, where the construct `tick` denotes the implicit clock.

Listing 11.3 Snippet of the *fir* component in Esterel.

```
module fir :
  extends FirIntf;
    every tick do
      if (?reset = true) then
        // Reset Behavior
      end if;
      if (?reset = false) and (?in_valid = true) then
        // Filter Computation
      end if
end module
```

These behaviors specified in the functional model are shown in Listing 11.4. The construct `emit` denotes a write on the output, which is available for reading in the next cycle. The *shift* is a local signal array that represents the tap registers of the FIR, whereas *prod* and *accu* are local signals that are employed in the *fir* computation to hold the product and accumulated result.

Listing 11.4 *fir* behavior in Esterel.

```
// Reset Behavior                  //Tap Shifting
for j < 16 dopar                   emit ?shift[1..15] <= pre(?shift[0..14]);
```

```
        emit ?shift[j] <= 0             emit ?shift[0] <= ?sample;
end for;                                //Convolution
emit ?result <= 0;                      for j < 16 dopar
emit ?out_data_rdy <= false;                emit ?prod[j] <= Coef[j] * ?shift[j]
                                        end for;
                                        emit ?accu <= ?prod[0] + ?prod[1] + ?prod[2]
                                                  + ?prod[3] + ........ + ?prod[15];
                                        //Notification
                                        emit ?result <= ?accu;
                                        emit ?out_data_rdy <= true;
```

We recommend certain modeling guidelines during the functional modeling in Esterel. These restrictions are: (1) user must specify the model imperatively as opposed to the alternate equational style (Listing 11.5) and (2) user must limit the usage of concurrent emissions and use ; for sequential emissions. These restrictions are needed to match up the functional model with the implementation, since one of the objectives is to provide code coverage on the implementation by executing the test generated from the functional model. The variations between the functional and implementation model that arise due to the differences in modeling languages are taken care of during the test generation and transformation.

Listing 11.5 Modeling illustrations for the restrictions.

```
Imperative                  |   Equational
if I then emit J end if     |   emit J <= I
```

Note the functional model is described such that the readers can see the structural resemblance to the implementation model in the SystemC distribution [3]. For further details on the modeling constructs in Esterel and its FSM semantics, we recommend [1,2].

Verification Model

In Stage 1, the user also derives behavioral properties from the English specification [4] and builds a verification model. These properties expressed in PSL are used to build nonintrusive verification monitors in Esterel, which sit on top and verify the components of the functional model and their interaction. In order to verify that the functional model preserves these properties, we execute the model with the monitors. If violations exist during simulation, the user is notified with VCDs and waveforms of the execution traces. Note that these properties can be asserted as part of the functional model and verified through the FV engine in ES, which is how typically ABV is performed. However, we build monitors first to decouple the verification process and second to convert these monitors into conformance tests to validate the implementation. Note that the property derivation in PSL and the verification model development are manually performed.

PSL [5] is used to express properties capturing the temporal relation between elements of the design such as inputs and outputs or reads and writes. It allows specifying assertions, which constrain the design's behavior based on the properties. For example, the reset behavior gives rise to interesting properties about the design. For the *fir* example, the reset property "`if reset is asserted in the first cycle, then outputs are de-asserted in the next cycle,`" is formulated textually in PSL and as an executable Esterel assertion in Listing 11.6.

Listing 11.6 Property assertion for *fir* in Esterel.

```
// PSL Property (RESET_HIGH)
assert ( reset -> next ( (result = 0) && !out_data_rdy) );

// Esterel monitoring assertion
assert RESET_HIGH =
  ( next(?result = 0) and next(?out_data_rdy = false) ) if (?reset = true)
```

PSL also allows specifying assumptions on the environment, which are expectations of the component regarding its neighbors. These component-level assumptions are asserted on the complete design when the component is forced to interact with its environment. It is important to specify the environmental assumptions on the subcomponents of a design, because it improves the quality of the test generated from these subcomponents in isolation. For example, how often is the reset expected, and how long will it be asserted are typical assumptions on the reset stimulus. For the *fir*, the reset property "`reset is only asserted in the first three cycles`" is formulated textually in PSL and as an executable Esterel assertion in Listing 11.7. This assertion can only be verified on a design that utilizes the *fir* as a subcomponent. It is flagged as an ***Assume*** for the *fir*, which configures the FV engine in ES to generate values for the *reset* taking the assumption into consideration during test generation.

Listing 11.7 Property assumption for *fir* in Esterel.

```
// PSL Property (RESET_THREE_CYCLES)
assume ( rose(reset) -> next_a[1:2] (reset) );

//Esterel monitoring assertion
abort
  assert RESET_THREE_CYCLES = (?reset = true)
after
  await 3 tick;
end abort
```

ES can also generate assertions automatically by statically analyzing the functional model. These are structural assertions that check for overflow conditions, read-before-write scenarios, and race conditions. The structural and behavioral assertions are initially used to verify the correctness of the functional model, but in Stage 2 they are instrumented and in Stage 3 utilized for test generation.

11.1.2 Coverage Annotation

In Stage 2 after completion of the functional and verification models, they are annotated with hints in the form of assertions and assumptions. These instrumented models are given as input to the FV engine for the purpose of test generation with certain flags set on the annotations. As in any typical verification flow, the FV engine constrains based on the assumptions and verifies the assertions. If any assertion fails, then the FV engine spits out a counterexample, that we call a *test input*, for debugging the model. This feature of the FV engine is exploited for the purpose of test generation. *In our instrumentation, we take a **verified** assertion and assert its negation, which is reverified through the FV engine. This compels the FV engine to generate a **counterexample** to test the original assertion on the functional model.*

For example, consider **p**, which is either a statement in the Esterel code or a property that is asserted and verified. The instrumentation performed is **assert not(p)**. When the FV engine fails to verify this annotation, it produces a countertest **t** for **assert not(p)**, which is a test for **p** or **assert p**. If **p** is an **always q** property, then its negation results in an eventuality property, which is verified using bounded model checking techniques. If a countertest cannot be generated within a specified bound, then this verified property cannot be validated on the implementation within that bound.

In Stage 3, this test is translated into a concrete test that verifies the assertion on the implementation model, therefore validating the implementation against the functional model. In this section, we discuss the instrumentation for the purpose of code and functional coverage.

Structural Coverage Model

The model obtained from the instrumentation of the functional model for code coverage on the implementation is called the *structural coverage model*. The code coverage metrics employed range from the traditional metrics such as statement and branch, to the complex metric MCDC [6]. Note that the annotations for code coverage are assertions formulated with the negation logic for the purpose of test generation.

Statement Coverage: We instrument this coverage by asserting the negation of every statement of the functional model, which boils down to asserting the emissions (*emit* & `sustain` statements) by replacing the assignment (<=) by an equality (=) operation. For example in the *fir*, we illustrate how to cover the shift operation through an assertion in Listing 11.8.

Listing 11.8 Instrumentation for statement coverage in *fir*.

```
// Statement in the functional model of fir
emit ?shift[1..15] <= pre(?shift[0..14]);

// Instrumentation for statement coverage
assert SHIFT_CODE = not ( ?shift[1..15] = pre(?shift[0..14]) );
```

Branch Coverage: A *condition* is defined as a Boolean expression with no Boolean operators. A *decision* is defined as a Boolean expression with zero or more Boolean operators. For example, a decision (A > 10 || (B = 1)) has two conditions (A > 10) and B.

We instrument branch/decision coverage by asserting the negation of the branch-taken and branch-not-taken for every branch statement in the functional model such as `if-else`, `case`, and `mux`. In the *fir*, we illustrate how to cover the *reset* branch in Listing 11.9, through the annotation of two parallel assertions.

Listing 11.9 Instrumentation for branch coverage in *fir*.

```
// Branch in the functional model of fir
if (?reset = true) then
   // Reset Behavior
end if

// Instrumentation for branch coverage
assert RESET_BRANCH_TRUE  = not (?reset = true)
||
assert RESET_BRANCH_FALSE = not (?reset = false)
```

We extend the branch coverage approach to obtain path coverage as shown in Figure 11.3. This is possible, because we track the various branch points through assertions and a Cartesian product of these assertions (a product of the branch points), gives us all the paths in the model. In order to avoid the path explosion problem, we consider a complex metric such as MCDC, which is very effective for control-dominated systems [7].

MCDC: We instrument MCDC coverage by asserting each condition (called `false-scenario`) as well as its negation (called `true-scenario`), such that if the condition is asserted, then so is the decision and vice versa. For the

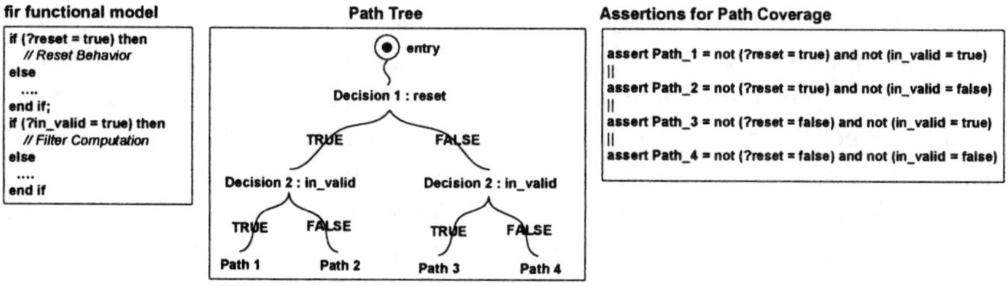

Figure 11.3 Path coverage through assertions.

`true-scenario` of a condition-decision pair, we assert the negation of the condition as well as the negation of the whole decision in the same assertion. When the FV engine is asked to verify this assertion, it sets this condition to true and sets all other ANDed/ORed conditions to avoid logical masking, such that the decision is also true. For the `false-scenario` of a condition-decision pair, we assert the condition as well as the whole decision in the assertion. Note that if all the conditions are ANDed together in a decision, then the `true-scenario` for each condition results in the same value assignment to the variables of the decision. Therefore, only one assertion needs to be specified to cover the `true-scenario` for all conditions. Similarly, for conditions ORed together in a decision, only one assertion needs to be specified to cover the `false-scenario`. In the *fir* model shown in Listing 11.3, we illustrate MCDC coverage on the branch statement that triggers the filter computation in Listing 11.10 through the annotation of three parallel assertions. For a decision with m conditions, MCDC requires $m + 1$ test cases, therefore it avoids the state explosion problem.

Listing 11.10 Instrumentation for MCDC coverage in *fir*.

```
// Branch in the functional model of fir
if (?reset = false) and (?in_valid = true) then
   // Filter Computation
end if

// Instrumentation for MCDC coverage
// C1: (?reset = false)
// C2: (?in_valid = true)
assert C1_FALSE_BR_FALSE = not(not(?reset = false) and (?in_valid = true))
||
assert C2_FALSE_BR_FALSE = not((?reset = false) and not(?in_valid = true))
||
assert C1/C2_TRUE_BR_TRUE = not((?reset = false) and (?in_valid = true))
```

The modeling restriction of an imperative specification allows the control flow in the functional model to be identical to that of the implementation. This enables achieving code coverage on the implementation using tests generated from the functional model. Note that currently we manually perform the instrumentation for the various code coverage metrics. However, the annotations can be automated through the compiler, because the assertions are generated in a very systematic manner, as described above. Furthermore, the assertions are specified in parallel to the model using the construct (||). This allows the assertions to be inserted at the end of the model, which eases the automated instrumentation process with very little alteration to the model.

Conformance Model

The model obtained from the instrumentation of the verification model for functional coverage of the implementation is called the *conformance coverage model.* We attain functional coverage by annotating the verification monitors, where we negate every **assert** property and leave the **assume** property untouched. These annotations require appending the `not` operator to every Esterel assertion that enforces an assert safety property. Therefore the automation can be easily achieved and is nonintrusive similar to the instrumentation for code coverage.

11.1.3 Test Generation

In Stage 3, the annotated assertions specified in the structural and conformance model are verified through *esVerify* after application of the appropriate assumption. All these assertions fail providing counterexamples in *.esi files* to trace the failure. Note that, the assertions are formulated with the negation logic, so that the counterexamples generated are tests for the original assertions. A counterexample (.esi file) is a spec describing a number of steps, with each step having value assignments for the inputs on the component interface. Consider the *RESET_BRANCH_FALSE* assertion specified for branch coverage of the *fir* in Listing 11.9, the counterexample generated by *esVerify* is shown in Listing 11.11. In this counterexample, each step refers to a clock cycle for the *fir* component.

Listing 11.11 Test inputs generated for *RESET_BRANCH_FALSE* coverage in *fir*.

```
% assertion "fir/RESET_BRANCH_FALSE" is violated at depth 0
%------------------------------
% BEGIN
% STEP: 0
  rst = false; input_valid = true; sample = 0;
% END
%------------------------------
```

The test example in Listing 11.11 specifies a set of input values of the *fir*, which satisfies the assertion without an environmental assumption. This may result in nonrealistic test inputs; mostly while component-level testing. For the *fir* example, we know that the *RESET_THREE_CYCLES* assumption exists; therefore, it is necessary to flag it as an ***Assume*** and then verify the previous assertion. This generates realistic test inputs as shown in Listing 11.12.

Listing 11.12 Test inputs generated with the assumption RESET_THREE_CYCLES.

```
% Assumed assertions: "fir/RESET_THREE_CYCLES"
% assertion "fir/RESET_BRANCH_FALSE" is violated at depth 3
%-----------------------------
% BEGIN
% STEP: 0
  rst = true; input_valid = true; sample = 0;
% STEP: 1
  input_valid = false;
% STEP: 2
  input_valid = true;
% STEP: 3
  rst = false;
% END
%-----------------------------
```

Note that the counterexamples are not complete tests, because the corresponding outputs for the inputs for a step are not generated as part of the .esi file. In order to obtain the corresponding outputs, we entail the code generation capability of ES in the next part of Stage 3.

Code Generation

We utilize the code generation capability of ES to transform the functional model into a C model and translate structural and conformance test inputs (.esi files) into testbenches in C. These testbenches are executed on the functional model and the generated outputs are recorded in *.eso files*. The APIs for capturing the outputs are also automatically code generated as a part of the testbench. Note that, our C-based approach of recording the outputs can be performed on one of the other targets as well. The outputs recorded for the test inputs (Listing 11.12) of the *RESET_BRANCH_FALSE* assertion in the *fir*, is shown in Listing 11.13.

Listing 11.13 Test outputs captured for *RESET_BRANCH_FALSE* coverage in *fir*.

```
% STEP: 0
  out_data_rdy = false; result = 0;
% STEP: 1
  out_data_rdy = false; result = 0;
% STEP: 2
  out_data_rdy = false; result = 0;
```

```
% STEP: 3
  out_data_rdy = true; result = -6;
```

An input scenario generated by *esVerify* and the outputs recorded by executing the input scenario on the functional model are combined together to form an *abstract test*, which is explained in the following section. The abstract test suite generated from the structural coverage model attains code coverage of the system implementation, whereas the suite generated from the conformance coverage model attains functional coverage of the system implementation.

11.2 SYSTEMC VALIDATION

An abstract test suite is a collection of test data and test cases having an abstract view on issues like communication, time, and data. Tests derived from the functional and verification model are by nature abstract. These abstract tests must be transformed into their concrete executable form in order to validate the implementation against the reference model.

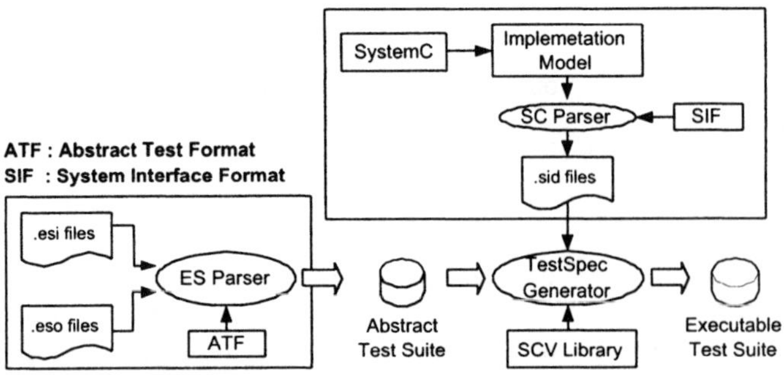

Figure 11.4 Executable test suite generation for SystemC validation.

Our transformation flow is shown in Figure 11.4, where the executable test suite generated is employed in validating the implementation model developed in SystemC. The abstract test suite is created from the input and output scenarios obtained from ES, where these scenarios are extracted from the .esi and .eso files using a parser that we implemented. The ES parser populates an abstract test format (ATF) in order to generate the different test cases. The ATF is an XML schema for describing an abstract test case, which captures the number of execution steps needed and the input and output assignments for each step. In Listing 11.14, we illustrate the test case obtained by populating the ATF for the RESET_BRANCH_FALSE scenarios in Listing 11.11.

Listing 11.14 The *RESET_BRANCH_FALSE* test specified using ATF.

```
<test_suite>
   <test_case name = "RESET_BRANCH_FALSE">
      <execution_steps length ="1">
         <step number = "1">
            <input_scenario>
               <input name = "reset"    value = "false"/>
               <input name = "in_valid" value = "true"/>
               <input name = "sample"   value = "0"/>
            </input_scenario>
            </output_scenario>
               <output name = "out_data_rdy" value = "true"/>
               <output name = "result"       value = "-6"/>
            </output_scenario>
         </step>
      <execution_steps>
   </test_case>
</test_suite>
```

The implementation model is also analyzed to extract the structural meta-information by a SystemC parser (SC Parser) that we implemented. The metadata describes the component characteristics that are utilized for exciting the component during testing, such as input-output ports for RTL components and read-write interfaces for transaction-level (TL) components. We also extract the sensitivity list of the component, configurable parameters as well as the module hierarchy, and module connectivity. The SC parser is provided with the system interface format (SIF) schema as input along with the implementation model. The parser populates the SIF with the extracted SystemC metadata and creates a .sid file. This file generated for the *fir* component is shown in Listing 11.15.

Listing 11.15 Structural metadata on the *fir* specified using SIF.

```
<system_model name ="fir_filter">
   <component name = "fir" abstraction="RTL">
      <clockport  datatype="bool" name="clk"/>
      <inputport  datatype="bool" name="reset"/>
      <inputport  datatype="bool" name="in_valid"/>
      <inputport  datatype="int"  name="sample"/>
      <outputport datatype="bool" name="out_data_rdy"/>
      <outputport datatype="int"  name="result"/>
      <sensitive_list>
        <port name = "clk" pos_edge = "true" neg_edge = "false"/>
```

```
        </sensitive_list>
    </component>
<system_model>
```

The SIF schema also has XML elements to capture transaction descriptions. In Listing 11.16, we illustrate the .sid file for the nonblocking TL interface of the *simple_bus* component from the SystemC distribution.

Listing 11.16 Structural metadata on the TL *simple_bus* [3] specified using SIF.

```
<system_model name = "simple_bus">
  <component name = "simple_bus_master_non_blocking" abstraction = "TL">
    <param name = "timeout"   datatype = "int" value = ""/>
    <param name = "lock"      datatype = "bool" value = ""/>
    <transaction_interface = "non_blocking_intf">
      <transaction type = "READ" implementation_method = "nb_read">
        <implementation_method name = "nb_read">
          <arg name = "data"    datatype = "int"  type = "OUTPUT"/>
          <arg name = "address" datatype = "int"  type = "INPUT"/>
          <arg name = "prior"   datatype = "int"  type = "INPUT"/>
          <arg name = "lock"    datatype = "bool" type = "INPUT"/>
        </implementation_method>
      </transaction>
    </transaction_interface>
  </component>
</system_model>
```

11.2.1 TestSpec Generator (TSG)

The TSG transforms the abstract test suite into an executable test suite. Inputs to the TSG are the abstract test suite and .sid file generated from the implementation model. The output is an executable SystemC test suite. The main objective of the transformation is to map the input output values of an abstract step into a simulation step, where the abstract step could be a cycle, multiple events, or a complete transaction. The mapping is guided by the .sid file that specifies whether the implementation is a sequential/combination block or a TL component with read-write interfaces.

Listing 11.17 Executable test for *RESET_BRANCH_FALSE* coverage in *fir*.

```
sc_module(RESET_BRANCH_FALSE_test)        void RESET_BRANCH_FALSE_test::main()
{                                         {
    sc_in<bool>   clk;                      while(true)
                                             {
    // Driving                                 drive_sample = 0;
    sc_in<int>  drive_sample;                  drive_reset = false;
    sc_in<bool> drive_in_valid;                drive_in_valid = true;
    sc_in<bool> drive_reset;
                                               drive_clk = 1;
    // Checking                                wait();
```

```
    sc_in<bool> rec_out_data_rdy;
    sc_in<int>  rec_result;

    // Clocking
    sc_out<bool> drive_clk;

  SC_CTOR(RESET_BRANCH_FALSE_test)
  {
    SC_THREAD(main);
    sensitive<<clk.pos();
  }
  void main();
};
```

```
      drive_clk = 0;
      wait();

      if(rec_out_data_rdy != true
         ||
         rec_result != -6)
       cerr<<"Test Failed"<<endl;
    }
  }
```

For a clock-driven component the mapping is straightforward, where every abstract step is one clock cycle in the implementation model. In this cycle, inputs of the abstract test drive the implementation. The implementation outputs are checked against that of the abstract test in the next cycle. In Listing 11.17, we illustrate the executable test generated from the abstract test in Listing 11.14, which attains *RESET_BRANCH_FALSE* coverage. For an event-driven component, an abstract step can be mapped to a single/multiple cycles, but outputs are instantaneously available for checking when inputs are applied. For a TL component, an abstract step can be mapped to a transaction read or write. The TSG employs the SystemC verification library (SCV) to randomize the testing of implementation aspects that are don't-cares in the abstract test. It also facilitates transaction recording and VCD dumping. Our executable test suite generation employs the ATF and SIF to provide flexibility in terms of the abstract test generation framework and the system-level design language. The executable tests generated have built-in checkers to notify any violation during the validation of the implementation model.

11.3 CASE STUDY

In this section, we illustrate our methodology through the case study of a Power State Machine (PSM).

11.3.1 PSM Specification

PSM is an essential ingredient in reducing the power consumption of the system by regulating the power states. It is distributed along the design; hence the validation of the PSM requires simulating almost the whole system. In Figure 11.5, we show the four different states of the PSM, namely `Active`, `Idle`, $\texttt{Sleep}_1$, and $\texttt{Sleep}_2$

and its four different constituents namely *Queue*, *Timer*, *Sampler*, and *Service Provider*.

- **PSM:** The system remains in `Active` state if it is receiving a task. If no task is received for t_1 clock cycles, then it goes to the `Idle` state. In the `Idle` state, if it does not receive any task for t_2 cycles, then it goes to the `Sleep`$_1$ state. In `Sleep`$_1$, if the system does not receive any task for t_3 cycles, then it goes to the deep sleep state `Sleep`$_2$. From any of these states, the system returns to the `Active` state if it receives a task.
- **Sampler:** All power states have a sampler that samples a task every t_4 cycles.
- **Queue:** A task received by the system is stored on a finite queue, which is accessible to every state.
- **Service Provider (SP):** In the `Active` state, the SP processes a sampled task for t_5 cycles. The SP requests for a task and receives a task that is dequeued.
- **Timer:** Every state except the `Sleep`$_2$ has a timer embedded, which tracks for how long the system is idle without receiving a task.

Validation of the PSM is very important because such a component is often used in system-level power management for reduction of power in embedded systems.

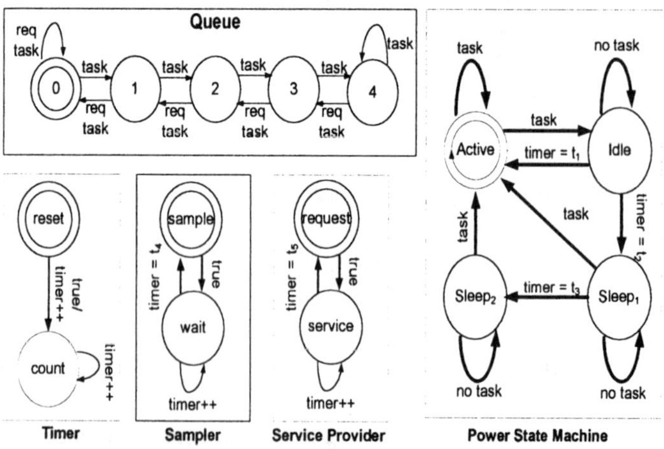

Figure 11.5 Specification of the different PSM ingredients.

11.3.2 Functional Model

The functional model was created in Esterel by strictly following the modeling paradigm discussed in Section 11.1.1. It is approximately 150 lines of code. The functional model is described such that the control flow and module boundaries are identical to the ones in implementation model discussed in Section 11.3.6. However, the way Esterel expresses hierarchy is different from SystemC, therefore certain internal states of the embedded components need to be exposed and tunneled for visibility at the toplevel. This results in additional ports at the module boundary, which are ignored during test generation. Furthermore, Esterel uses implicit connections between communicating components that drastically reduce the code size, but introduce new wrapper modules. The tests are only generated on components that exist in the implementation model.

11.3.3 Verification Model

We derived about 90 properties from the PSM specification and expressed them in PSL. These properties are specified on the individual components such as Queue and Timer, as well as on the interaction of the different components such as the Sampler-Timer and PSM-Timer. In Table 11.1, we enumerate the number of properties derived and expressed in terms of assertions and assumptions on the different subcomponents and their interaction. These properties were modeled as verification monitors in Esterel using assertions and were used to verify the correctness of the functional model. In the following stage of the methodology, these assertions were instrumented to be leveraged for test generation.

Table 11.1

Number of Properties Derived from the Specification

Property	Toplevel	PSM	Sampler	Queue	Timer
Assertion	18	27	23	12	9
Assumption	1	1	1	1	1

11.3.4 Coverage Annotation

Table 11.2 shows the number of annotations made in the functional as well as the implementation model. For statement coverage, we annotate every `emit` and `sustain` statement, since they describe emissions that change the state of the PSM. The branch statements such as `if-then-else`, and `case` are covered during branch coverage. MCDC is a complex metric used to cover decisions with multiple conditions, but this model does not have many such decisions. We also instrument the monitors in the verification model with assertions.

Table 11.2
Number of Annotations for Coverage Instrumentation

Component	Structural			Conformance
	Statement	Branch	MCDC	MONITOR
Toplevel	31	52	54	18
PSM	28	44	45	27
Sampler	13	22	23	23
Queue	3	8	9	12
Timer	13	12	13	9

11.3.5 Test Generation

The instrumented models are verified through *esVerify* to generate counterexamples, which are transformed into executable tests using our integrated framework. The total number of test cases generated for our case study is shown in Table 11.3.

Table 11.3
Executable Test Suite

	Structural Test	Conformance Test
Test suite	370	89

11.3.6 Implementation Model

The PSM is implemented as a hierarchical model in SystemC using approximately 500 lines of code. Its components are described as `SC_THREAD`-based processes. The implementation details are briefly outlined below:

- **Sampler:** has an input that allows sampling and an input to receive the task. It has a *Timer* embedded to control the sampling interval. It has two outputs that indicate if a task needs to be queued or if an SP request needs to be processed.
- **Timer:** has a reset input and an input to indicate when to begin the timer. It has an output to indicate when the *Timer* is done. It is reused at five different places in this model with different threshold values.
- **Queue:** serves as a finite memory component with two inputs, one of which indicates that a task needs to be queued and the other indicates that a task needs to be dequeued and given as output to the SP.
- **PSM:** is a clocked component that models and switches between the different power states. It embeds three different *Timers* to determine how long to stay in these states. It receives a task as input that is given to the embedded *Sampler*. Outputs of the *Sampler* are the outputs of the PSM.
- **Toplevel:** models the interaction between the PSM and the *Queue*. Tasks obtained from the *Queue* are eventually given to the SP.

11.4 SUMMARY

In this chapter, we presented an Esterel-based methodology for model-driven validation of SystemC designs through test generation. The test generation is performed through coverage instrumentation and by using the countertrace generation capability of Esterel Studio. The input and output test scenarios generated through ES are abstract tests used to attain structural and functional coverage of the implementation. These are transformed into executable SystemC tests through a TestSpec generator and a system interface description of the implementation model. Our methodology requires a good knowledge of the Esterel language and the verification capabilities of Esterel Studio. Furthermore, the countertrace generation limits the test generation to a single test solution even if there are multiple ways to test a particular implementation aspect. In our future work, we will perform empirical evaluation of the bug-finding capability of different coverage metrics as well as the stability of different coverage metrics across test suites.

References

[1] G. Berry, "The Foundations of ESTEREL," *Proof, Language and Interaction: Essays in Honour of Robin Milner*, 2000, pp. 425–454.

[2] Esterel Technologies Inc., "Esterel Studio," http://www.esterel-technologies.com/products/esterel-studio/.

[3] OSCI Group, SystemC Web site, http://www.systemc.org/.

[4] S. Suhaib, D. A. Mathaikutty, S. K. Shukla, and D. Berner, "XFM: An Incremental Methodology for Developing Formal Models," *ACM Transaction on Design Automaton of Electronic Systems*, Vol. 10, No. 4, 2005, pp. 589–609.

[5] PSL/Sugar Consortium, "Property Specification Language," http://www.pslsugar.org/.

[6] D. A. Mathaikutty, S. V. Kodakara, A. Dingankar, S. K. Shukla, and D. Lilja, "Design Fault Directed Test Generation for Microprocessor Validation," *Proceedings of Design Automation and Test Conference Europe (DATE'07)*, 2007, pp. 761–766.

[7] A. Dupuy and N. Leveson, "An Empirical Evaluation of the MC/DC Coverage Criterion on the HETE-2 Satellite Software," *Proceedings of the Digital Aviation Systems Conference (DASC'02)*, October 2000.

Chapter 12

Conclusion and Future Work

12.1 SUMMARY

This book addresses two important problems in reusing intellectual properties (IPs) in the form of reusable design components, or reusable verification components. The first problem is associated with fast and effective integration of reusable design components into a system-on-chip (SoC). The second problem emphasizes verification model reuse, where it specifically addresses reuse of reusable verification IPs to enable a "write once, use many times" verification strategy. This book is accordingly divided into Part I and Part II, which are related but describe the two problems and our solutions to them. Our solutions employ a metamodeling technique that is commonly used in the software engineering domain to create customized metalanguage to describe the syntax and semantics for a modeling domain. We extend this technique with various innovative modifications to meet our goals for SoC and microprocessor design. The acceptance of this work in conferences [1–14], journals [15–20], and book chapters [21–23] leads us to hope that these solutions and their enabling technologies we proposed are gaining acceptance.

In Part I, we considered the problem of rapid system-level model integration of existing reusable IPs and addressed the following questions: (1) how to formally specify the SoC architectural requirements, (2) how to automatically select IPs from an IP library to match the need of the system being integrated, (3) how to prove or automatically check the integrability of the IPs, (4) how to identify, extract, and represent structural and behavioral type systems for each IP, and (5) how to automatically generate alternative system-level models. Our solutions consist of a component composition language, algorithms for IP selection, type matching and inferencing algorithms, type checking algorithms, temporal property based behavioral typing, and finally a software system on top of an existing metamodeling environment.

We developed *MCF*, a metamodeling-driven component composition framework to enable IP-reuse for fast SoC integration. It has three main ingredients: (1) modeling environment [1, 16, 21, 24, 25], (2) IP reflector, selector, and restricter [2,3,19,22], and (3) interaction console [5]. The modeling environment [16] facilitates the creation of an architectural specification of the system in terms of abstract modules, ports, interfaces, buses, and so forth. It also allows a property specification to annotate these blocks with some behavior. To enable a rapid specification and alleviate some of the software engineering issues, the environment allows the designer to create descriptions with partial/incomplete components. The environment applies automated inference techniques to derive possible interfaces of partially specified blocks and performs design-time checks to avoid modeling violations. The architectural template is used to guide the automated searches for various implementations from an IP library, so that the entire template can be instantiated by substituting abstract modules with real implementations from the IP library and create executable models. Our library is a collection of compiled SystemC IP-cores from the SystemC distribution [26] and industry contribution. The IP reflector extracts structural and composition metadata from the IPs in the library [2,19]. The extracted metadata is used to perform automated abstraction and design visualization-based matching between the virtual components in the architectural template of the system and the implementations in the IP library [3, 19]. The IP restricter allows the addition of generic IPs to the library and facilitates extraction, selection, and composition of these generic IPs. The interaction console allows the designer to guide the IP integration process [5] in terms of the preferred selection modes, ideal configurations, and so forth.

We also formulate the type inference problem that exists in MCF and discuss our multistage solution that performs type-propagation and type-matching to convert a partially typed architectural specification into a completely typed one [4]. The three requirements to solve the MCF type inference problem are: (1) the architectural specification, (2) the IP library, and (3) the MCF type-system. The MCF type-system captures the relevant types of the target design language and their relationships. The type-system is given as input to the inference techniques for the purpose of type-matching. These techniques consult the type-system during type-checking to determine whether there is a type mismatch in the specification. The MCF type inference is a two-part problem, where the first part addresses the type inference issues at the architectural level with abstract components and the second part addresses the issues with replacing abstract components in the template with real implementations from an IP library. In this section, we formally define the terminologies necessary to comprehend the MCF type inference problem.

Note that MCF was built as a proof-of-concept to demonstrate SoC integration by employing reusable SystemC IPs and an architectural specification. To assess the scalability and usability of MCF, we have built a wide variety of models. These are examples from the SystemC distribution and designs obtained from our industrial contacts such as Intel. Some of the reasonably sized designs modeled are a microarchitecture of a RISC CPU with MMX capability [26] and a Digital Equalizer System [24], which are downloadable from [24].

In Part II, we refocused our attention on using the metamodeling technique to enable reuse of generative verification IPs. This part provided a verification strategy for Intel that spans the various abstraction levels (system, architectural, microcode, and microarchitecture). Our solution addresses the following questions: (1) how do we formally specify the relevant parts of the processor, (2) how do we capture the various modeling abstractions uniformly and enforce a well-defined refinement flow across abstractions, and (3) what are the verification collateral that need to be generated from the processor models and how are they automatically generated? Our solution provides a unified language that can be used to model verification IPs at each abstraction level, and various verification collaterals such as test benches, simulator plug-ins, coverage monitors, or temporal properties for model checking, can be generated from these generative IPs, thereby enhancing reuse in verification.

We developed *MMV*, a metamaodeling-driven modeling and validation environment to enable generative verification IP reuse for validating microprocessors [6, 7, 17]. MMV enables retargetable code generation and facilitates creating models at different levels of abstraction, without tying them to any specific programming language. The language-independent representation enables the analysis and transformation of the models uniformly into various validation collaterals. It also allows a manual refinement methodology and enforces consistency within and across abstraction levels. We also illustrate the translation of instruction descriptions in the architectural model into executable description in C++ called ALGS. Furthermore, we show how the target and programming specifics are used to adapt the simulation model for next generation architectural requirements. As an example, we illustrate the modeling of Vespa [27], a 32-bit RISC processor at the system level, as well as architecture and microarchitecture levels to illustrate MMV's modeling capability and discuss how to extract constraints for test generation at each of these levels of abstraction.

The test generation framework [17] is extended to perform coverage-directed as well as design fault-directed test generation. The framework generates test suites that satisfy MCDC [28], a structural coverage metric that detects most of the classified design faults as well as the remaining faults not covered by MCDC [8–10]. We show that there exists good correlation between types of design faults proposed by software validation [29] and the errors/bugs reported in case studies on microprocessor validation. We demonstrate the framework by modeling and generating tests for the microarchitecture of Vespa and show that the tests generated using our framework's coverage directed approach detects the fault classes with 100% coverage, when compared to pseudorandom test generation.

Finally, we developed a model-driven test generation framework that further enhances the verification experience through protocol checking and functional coverage metrics. The framework integrates Esterel [30] and SystemC for model-driven development and validation of system-level designs. The desired system is developed (functional modeling, simulation, and verification) using our design paradigm in Esterel Studio [31]. The test generation requires instrumentation and tests generated attain structural and functional coverage of the implementation. The structural coverage attained range from traditional metrics such as statement and branch, to a complex metric such as modified condition/decision coverage (MCDC). The functional coverage is attained using temporal properties specified in the property specification language (PSL) [32]. Finally, we transform the abstract tests generated from Esterel Studio into executable tests through a TestSpec generator for validating the system implementation in SystemC [26].

In summary, the theme of this book is reuse of intellectual property, contexts are system-level modeling for design space exploration and verification of microprocessors, and the solution strategy is metamodeling techniques.

12.2 FUTURE WORK

There are several opportunities for enhancing the work presented in this book. They can be divided into two categories, namely solution methodology enhancements and extensions for industry recognition.

12.2.1 Solution Methodology Enhancements

Methodology enhancement to Part I of the book will include: (1) behavioral typing to see how compatible two IPs are, and what is needed to substitute one with the other, (2) support for subtyping relations during type inference, which will widen the solution space to solve the IP composition problem, (3) type-lattice construction from the hardware domain types with the top element being the most general type and the bottom element being UN to improve the robustness of the type inference algorithms, (4) behavioral type-system extraction and representation, (5) automatic adapter synthesis for partially matching IPs obtained during IP selection to increase the number of possible matches, and (6) enable reuse through behavioral type-theoretic techniques that rely on the behavioral type-system to enhance the correctness of IP integration.

Methodology enhancement to Part II of the book will include: (1) test generation framework to perform exception coverage, event coverage, and sequencing coverage for instructions, (2) performing empirical evaluation of the bug-finding capability of different coverage metrics as well as the stability of different coverage metrics across test suites, and (3) analyzing multiple faults and coupling effect in the Boolean expression.

As the next step to the Esterel-based model-driven test generation work, we will automate the translation of cycle-accurate and TL SystemC models into Esterel models. The translation of a cycle-accurate model is straightforward and detailed in Chapter 11. However, the translation of a TL model is more difficult and would require creating an elaborate scheduler within the Esterel model to implement a multicycle synchronous model that mimics the TL model. Furthermore, we will develop the design methodology shown in Figure 12.1 to allow SoC development through the integrated framework presented in Chapter 11.

12.2.2 Extensions for Industry Recognition

A lot of industrial effort has gone into defining IP interfacing standards such as VSIA [33], OCP [34], and SPIRIT [35] to enable design reuse. Currently, effort is being put into developing tools and methodologies for IP reuse. For MCF developed in Part I of this book to be successfully adopted in the industry, it should be able to interoperate with existing solutions and must have supplementary features. We will equip the component composition language with the OCP protocol [34], which would enable expressing a large variety of hardware communication behaviors that promotes reusability at different abstraction levels within RTL and TL. For interoperability of MCF and SPIRIT-based tools [35], we will develop a generic metamodel that imports the SPIRIT metadata and implements the respective APIs as plug-ins to facilitate the IP integration as part of MCF. Furthermore, we will utilize SPIRIT schemas for capture of IP metadata, which would enable importing IPs represented in SPIRIT into our modeling framework as component libraries. These libraries will be employed during the creation of the architectural template.

Furthermore, the IP libraries could be expanded to hold IPs specified in other ESL languages such as Esterel [30], SystemVerilog [36], or any HDL-based on C/C++. Finally, MCF will also be used in creating a power-estimation methodology, where metamodels for system-level and the RTL models will be captured and the various correspondences between the model variables will be generated. It will be utilized to express metainformation about RTL and system-level models, and how model transformation rules can be used to capture mapping between SLD variables and RTL variables, which will then be used to algorithmically derive the activity statistics at a lower level from traces in the higher-level model executions.

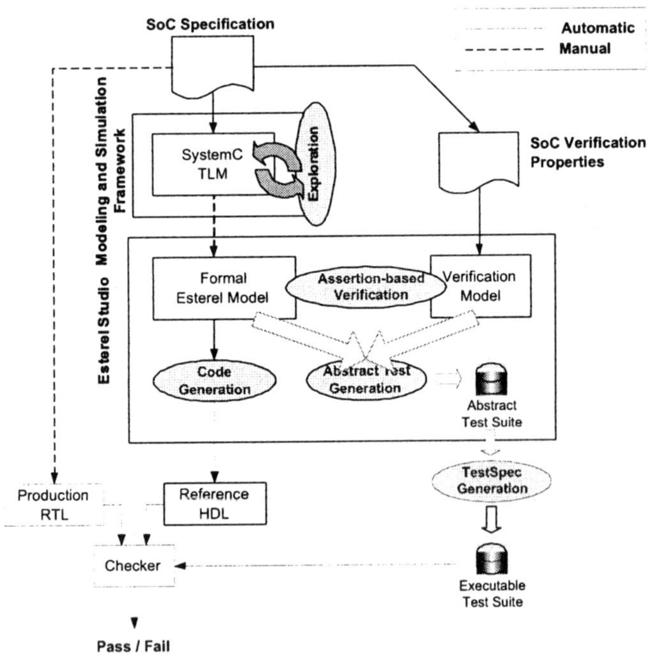

Figure 12.1 SoC development flow.

The additions to Part II of the book for industrial acceptance will focus on developing MMV into a full-fledge validation environment for Intel's microprocessors. We will extend the environment for creating other verification collaterals, namely software simulation, hardware synthesis for emulation, and other coverage analysis. Currently, the simulator generation employs Intel instruction set APIs for code generation. In future work, we will automatically generate the states of the processor as well as the necessary APIs from the architectural model. Finally, some extensions that focus on the tool integration and methodology development will also be done.

References

[1] D. Mathaikutty and S. Shukla, "MCF: A Metamodeling Based Visual Component Composition Framework," *Proceedings of Forum on Specification and Design Languages (FDL'06)*, TU Darmstadt, September 2006.

[2] D. Mathaikutty and S. Shukla, "Mining Metadata for Composability of IPs from SystemC IP Library," *Proceedings of Forum on Specification and Design Languages (FDL'06)*, TU Darmstadt, September 2006.

[3] D. Mathaikutty and S. Shukla, "SoC Design Space Exploration Through Automated IP Selection from SystemC IP Library," *Proceedings of IEEE International SoC Conference (SOCC'06)*, Austin, TX, September 2006, pp. 109–110.

[4] D. Mathaikutty and S. Shukla, "Type Inference for Component Composition," *Proceedings of 5th ACM-IEEE International Conference on Formal Methods and Models for Codesign (MEMOCODE'07)*, May 2007, pp. 61–70.

[5] D. Mathaikutty and S. Shukla, "On the Composition of IPs for Automated System Model Construction," *Proceedings of Technology and Talent for the 21st Century (TECHCON'07)*, Austin, TX, September 2007.

[6] A. Dingankar, D. Mathaikutty, S. Kodakara, S. Shukla, and D. Lilja, "MMV: Metamodeling Based Microprocessor Validation Environment," *Proceedings of IEEE International High Level Design Validation and Test Workshop (HLDVT'06)*, November 2006, pp. 143–148.

[7] D. A. Mathaikutty, A. Dingankar, and S. K. Shukla, "A Metamodeling Based Framework for Architectural Modeling and Simulator Generation," *Proceedings of Forum on Specification and Design Languages (FDL'07)*, Barcelona, September 2007.

[8] S. V. Kodakara, D. A. Mathaikutty, A. Dingankar, S. K. Shukla, and D. Lilja, "Model Based Test Generation for Microprocessor Architecture Validation," *Proceedings of the 20th International Conference on VLSI Design*, January 2007, pp. 465–472.

[9] D. Mathaikutty, S. Kodakara, A. Dingankar, S. Shukla, and D. Lilja,"Design Fault Directed Test Generation for Microprocessor Validation," *Proceedings of Design Automation and Test Conference Europe (DATE'07)*, March 2007, pp. 761–766.

[10] S. V. Kodakara, D. A. Mathaikutty, A. Dingankar, S. K. Shukla, and D. Lilja, "A Probabilistic Analysis For Fault Detectability of Code Coverage Metrics," *Proceedings of the 7th International Workshop on Microprocessor Test and Verification (MTV'06)*, December 2006.

[11] D. A. Mathaikutty, S. Ahuja, A. Dingankar, and S. K. Shukla, "Model-Driven Test Generation for System Level Validation," *Proceedings of IEEE International Workshop on High Level Design Validation and Test (HLDVT07)*, November 2007.

[12] D. A. Mathaikutty, H. D. Patel, and S. K. Shukla, "UMoC++: Modeling Environment for Heterogeneous Systems Based on Generic MoCs," *Proceedings of Forum on Specification and Design Languages (FDL'05)*, Lausanne, Switzerland, September 2005.

[13] D. Berner, H. D. Patel, D. A. Mathaikutty, and S. K. Shukla, "SystemCXML: An Extensible Systemc Frontend Using XML," *Proceedings of Forum on Design and Specification Languages (FDL'05)*, September 2005.

[14] D. Berner, H. D. Patel, D. A. Mathaikutty, and S. K. Shukla, "Automated Extraction of Structural Information from SystemC-Based IP for Validation," *Proceedings of the 6th International Workshop on Microprocessor Test and Verification (MTV'05)*, November 2005, pp. 99–104.

[15] D. Mathaikutty, H. Patel, S. Shukla, and A. Jantsch, "EWD: A Metamodeling Driven Customizable Multi-MoC System Modeling Framework," *ACM Transactions on Design Automation of Electronic Systems*, Vol. 12(3), No. 33, 2007, pp. 1–43.

[16] D. Mathaikutty and S. Shukla, "MCF: A Metamodeling Based Component Composition Framework — Composing SystemC IPs for Executable System Models," *IEEE Transactions on Very Large Integration (VLSI) Systems*, Vol. 16, No. 7, July 2008, pp. 792–805.

[17] D. Mathaikutty, S. Kodakara, A. Dingankar, S. Shukla, and D. Lilja, "MMV: Metamodeling Based Microprocessor Validation Environment," *IEEE Transactions on Very Large Integration (VLSI) Systems*, Vol. 16, No. 4, April 2008, pp. 339–352.

[18] H. D. Patel, D. A. Mathaikutty, D. Berner, and S. K. Shukla, "CARH: A Service Oriented Architecture for Validating System Level Designs," *IEEE Transactions on Computer-Aided Design of Integrated Circuits and Systems*, Vol. 25, No. 8, August 2006, pp. 1458–1474.

[19] D. Mathaikutty and S. Shukla, "Mining Metadata for Composability of IPs from SystemC IP Library, "*Design Automation for Embedded Systems*, Vol. 12, No. 1-2, 2008, pp. 63–94.

[20] D. Mathaikutty, H. Patel, S. Shukla, and A. Jantsch, "SML-Sys: A Functional Framework with Multiple Models of Computation for Modeling Heterogeneous System," *Design Automation for Embedded Systems*, Vol. 12, No. 1-2, 2008, pp. 1–30.

[21] D. Mathaikutty and S. Shukla, "MCF: A Metamodeling Based Visual Component Composition Framework," *Advances in Design and Specification Languages for Embedded Systems, Selected Contributions from FDL'06, Chapter 19*, 2007.

[22] D. Mathaikutty and S. Shukla, "Mining Metadata for Composability of IPs from SystemC IP Library," *Advances in Design and Specification Languages for Embedded Systems, Selected Contributions from FDL'06, Chapter 7*, 2007.

[23] D. A. Mathaikutty, H. D. Patel, and S. K. Shukla, "UMoC++: Modeling Environment for Heterogeneous Systems Based on Generic MoCs," *Advances in Design and Specification Languages for SoCs, Selected Contributions from FDL'05, Chapter 7*, 2006.

[24] FERMAT Group, MCF Web site, http://fermat.ece.vt.edu/MCF/MCF.htm, 2005.

[25] D. Mathaikutty and S. Shukla, "A Metamodel for Component Composition," FERMAT Lab., Virginia Tech, Tech. Rep. 2005-13, December 2005.

[26] OSCI Group, SystemC Web site, http://www.systemc.org/.

[27] D. J. Lilja and S. S. Sapatnekar, *Designing Digital Comptuer Systems with Verilog*, Cambridge, U.K.: Cambridge University Press, 2005.

[28] A. Dupuy and N. Leveson, "An Empirical Evaluation of the MC/DC Coverage Criterion on the HETE-2 Satellite Software," *Proceedings of the Digital Aviation Systems Conference (DASC'02)*, October 2000.

[29] M. F. Lau and Y. T. Yu, "An Extended Fault Class Hierarchy for Specification-Based Testing," *ACM Transactions on Software Engineering and Methodology*, Vol. 14, No. 3, 2005, pp. 247–276.

[30] G. Berry, "The Foundations of ESTEREL," *Proof, Language and Interaction: Essays in Honour of Robin Milner*, 2000, pp. 425–454.

[31] Esterel Technologies Inc., "Esterel Studio," http://www.esterel-technologies.com/products/esterel-studio/.

[32] PSL/Sugar Consortium, "Property Specification Language," http://www.pslsugar.org/.

[33] Virtual Socket Interface Alliance, http://www.vsi.org.

[34] OCP International Partnership Association, Inc, http://www.OCPIP.org.

[35] Spirit Consortium, "SPIRIT Schema Working Group," http://www.spiritconsortium.org/.

[36] System Verilog, http://www.systemverilog.org/.

About the Authors

Deepak A. Mathaikutty received a B.S. from the National Institute of Technology, Trichy, India, in 2003, and an M.S. and a Ph.D. in computer engineering from Virginia Tech, Blacksburg, in 2005 and 2007, respectively. He is currently a research scientist with the Microarchitecture Research Lab (MRL) of Intel Corporation, working on emulation platforms and technologies and microarchitectural exploration for lower-power IA (LPIA) design. His research interests include ESL modeling, formal verification, and dynamic validation of CPU and SoC. His current work includes metamodeling-driven design and validation of system-level models for microarchitectures and SoCs and a customizable multi-MoC modeling environment for system design. He has published 45 articles in journals and conference proceedings. His work was supported by the National Science Foundation (NSF), Semiconductor Research Corporation (SRC), and Intel Corporation.

Sandeep K. Shukla is an associate professor of computer engineering at Virginia Tech. He is also a founder and deputy director of the Center for Embedded Systems for Critical Applications (CESCA) and the director of the FERMAT Research Lab. He has published more than 125 articles in journals, books, and conference proceedings. He coauthored and coedited five books. He was awarded the Presidential Early Career Award for Scientists and Engineers (PECASE) for his research in design automation for embedded systems design, which in particular focuses on system level design languages, formal methods, formal specification languages, probabilistic modeling and model checking, dynamic power management, application of stochastic models and model analysis tools for fault-tolerant nanoscale system design, and reliability measurement of fault-tolerant nanosystems. Professor Shukla has been elected a College of Engineering Faculty Fellow at Virginia Tech and is a senior member of the IEEE and the ACM. He is a Wilheim Freidrich Bessel Fellow of the Humboldt Foundation. He has chaired a number of international conferences and workshops and edited over 10 special issues for various journals. He is on the editorial boards of *IEEE Design and Test of Computers*, *IEEE Transactions on Industrial Informatics*, *IEEE Transactions on Computer*, and *IEEE Embedded Systems Letters*. Currently he is working on model-driven embedded software synthesis from formal specifications.

Index

Recent Related Artech House Titles

Advanced Production Testing of RF, SoC, and SiP Devices, Joe Kelly and Michael Engelhardt

Design and Test of Digital Circuits by Quantum-Dot Cellular Automata, Fabrizio Lombardi and Jing Huang

Hybrid CMOS Single-Electron-Transistor Device and Circuit Design, Santanu Mahapatra and Adrian Mihai Ionescu

Integrated Interconnect Technologies for 3D Nanoelectronic Systems, Muhannad S. Bakir and James D. Meindl, editors

Introduction to Microelectromechanical (MEM) Microwave Systems, Hector J. De Los Santos

An Introduction to Microelectromechanical Systems Engineering, Second Edition, Nadim Maluf and Kirt Williams

Introduction to Self-Assembled Computer Design, Christopher Dwyer and Alvin Lebeck

Nanoelectronics Principles and Devices, Second Edition, Mircea Dragoman and Daniela Dragoman

Nano-Optics and Near-Field Optical Microscopy, Anatoly Zayats and David Richards, Editors

Post-Processing Techniques for Integrated MEMS, Sherif Sedky

Production Testing of RF and System-on-a-Chip Devices for Wireless Communications, Keith B. Schaub and Joe Kelly

RF MEMS Circuit Design for Wireless Communications, Hector J. De Los Santos

For further information on these and other Artech House titles, including previously considered out-of-print books now available through our In-Print-Forever® (IPF®) program, contact:

Artech House
685 Canton Street
Norwood, MA 02062
Phone: 781-769-9750
Fax: 781-769-6334
e-mail: artech@artechhouse.com

Artech House
16 Sussex Street
London SW1V 4RW UK
Phone: +44 (0)20 7596-8750
Fax: +44 (0)20 7630-0166
e-mail: artech-uk@artechhouse.com

Find us on the World Wide Web at: www.artechhouse.com